2024

IWUIC®

INTERNATIONAL WILDLAND-URBAN INTERFACE CODE®

ICC
INTERNATIONAL CODE COUNCIL®

2024 International Wildland-Urban Interface Code®

First Printing: August 2023
Second Printing: June 2024

ISBN: 978-1-959851-93-6 (soft-cover edition)
ISBN: 978-1-959851-94-3 (PDF download)

T032915

PRINTED IN THE USA

NEW DESIGN FOR THE 2024 INTERNATIONAL CODES

ICC INTERNATIONAL CODE COUNCIL®

IBC IRC IFC IPC IMC IECC IEBC IFGC IPMC IPSDC IWUIC IZC ICCPC IgCC ISPSC

The 2024 International Codes® (I-Codes®) have undergone substantial formatting changes as part of the digital transformation strategy of the International Code Council® (ICC®) to improve the user experience. The resulting product better aligns the print and PDF versions of the I-Codes with the ICC's Digital Codes® content.

The changes, promoting a cleaner, more modern look and enhancing readability and sustainability, include:

Streamlined lists

QR codes to identify code changes more accurately
(For further details, see Formatting Changes to the 2024 International Codes.)

Consistent grouping of associated content
(e.g., tables immediately follow parent sections)

2024

Single-column text

Shading for table headers and notes

Modernized font styles

ABCD
abcd

More information can be found at iccsafe.org/design-updates.

PREFACE

FORMATTING CHANGES TO THE 2024 INTERNATIONAL CODES

The 2024 International Codes® (I-Codes®) have undergone substantial formatting changes as part of the digital transformation strategy of the International Code Council® (ICC®) to improve the user experience. The resulting product better aligns the print and PDF versions of the I-Codes with the ICC's Digital Code content. Additional information can be found at iccsafe.org/design-updates.

Replacement of Marginal Markings with QR Codes

Through 2021, print editions of the I-Codes identified technical changes from prior code cycles with marginal markings [solid vertical lines for new text, deletion arrows (➡), asterisks for relocations (⋆)]. The 2024 I-Code print editions replace the marginal markings with QR codes to identify code changes more precisely.

A QR code is placed at the beginning of any section that has undergone technical revision. If there is no QR code, there are no technical changes to that section.

In the following example from the 2024 *International Wildland-Urban Interface Code*® (IWUIC®), a QR code indicates there are changes to Section 101 from the 2021 IWUIC. Note that the change may occur in the main section or in one or more subsections of the main section.

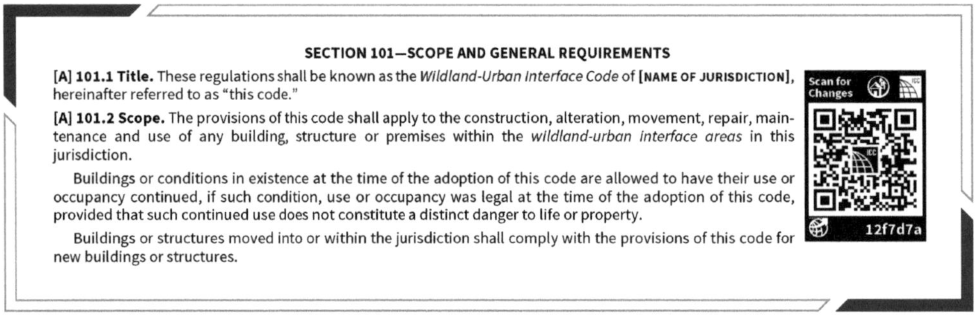
SECTION 101—SCOPE AND GENERAL REQUIREMENTS

[A] 101.1 Title. These regulations shall be known as the *Wildland-Urban Interface Code* of **[NAME OF JURISDICTION]**, hereinafter referred to as "this code."

[A] 101.2 Scope. The provisions of this code shall apply to the construction, alteration, movement, repair, maintenance and use of any building, structure or premises within the *wildland-urban interface areas* in this jurisdiction.

Buildings or conditions in existence at the time of the adoption of this code are allowed to have their use or occupancy continued, if such condition, use or occupancy was legal at the time of the adoption of this code, provided that such continued use does not constitute a distinct danger to life or property.

Buildings or structures moved into or within the jurisdiction shall comply with the provisions of this code for new buildings or structures.

To see the code changes, the user need only scan the QR code with a smart device. If scanning a QR code is not an option, changes can be accessed by entering the 7-digit code beneath the QR code at the end of the following URL: qr.iccsafe.org/ (in the above example, "qr.iccsafe.org/12f7d7a"). Those viewing the code book via PDF can click on the QR code.

All methods take the user to the appropriate section on ICC's Digital Codes website, where technical changes from the prior cycle can be viewed. Digital Codes Premium subscribers who are logged in will be automatically directed to the Premium view. All other users will be directed to the Digital Codes Basic free view. Both views show new code language in blue text along with deletion arrows for deleted text and relocation markers for relocated text.

Digital Codes Premium offers additional ways to enhance code compliance research, including revision histories, commentary by code experts and an advanced search function. A full list of features can be found at codes.iccsafe.org/premium-features.

ABOUT THE I-CODES

The 2024 I-Codes, published by the ICC, are 15 fully compatible titles intended to establish provisions that adequately protect public health, safety and welfare; that do not unnecessarily increase construction costs; that do not restrict the use of new materials, products or methods of construction; and that do not give preferential treatment to particular types or classes of materials, products or methods of construction.

The I-Codes are updated on a 3-year cycle to allow for new construction methods and technologies to be incorporated into the codes. Alternative materials, designs and methods not specifically addressed in the I-Code can be approved by the building official where the proposed materials, designs or methods comply with the intent of the provisions of the code.

The I-Codes are used as the basis of laws and regulations in communities across the US and in other countries. They are also used in a variety of nonregulatory settings, including:

- Voluntary compliance programs.
- The insurance industry.
- Certification and credentialing for building design, construction and safety professionals.
- Certification of building and construction-related products.
- Facilities management.
- "Best practices" benchmarks for designers and builders.
- College, university and professional school textbooks and curricula.
- Reference works related to building design and construction.

Code Development Process

The code development process regularly provides an international forum for building professionals to discuss requirements for building design, construction methods, safety, performance, technological advances and new products. Proposed changes to the I-Codes, submitted by code enforcement officials, industry representatives, design professionals and other interested parties are deliberated through an open code development process in which all interested and affected parties may participate.

Openness, transparency, balance, due process and consensus are the guiding principles of both the ICC Code Development Process and OMB Circular A-119, which governs the federal government's use of private-sector standards. The ICC process is open to anyone without cost. Remote participation is available through cdpAccess®, the ICC's cloud-based app.

In order to ensure that organizations with a direct and material interest in the codes have a voice in the process, the ICC has developed partnerships with key industry segments that support the ICC's important public safety mission. Some code development committee members were nominated by the following industry partners and approved by the ICC Board:

- American Gas Association (AGA)
- American Institute of Architects (AIA)
- American Society of Plumbing Engineers (ASPE)
- International Association of Fire Chiefs (IAFC)
- National Association of Home Builders (NAHB)
- National Association of State Fire Marshals (NASFM)
- National Council of Structural Engineers Association (NCSEA)
- National Multifamily Housing Council (NMHC)
- Plumbing Heating and Cooling Contractors (PHCC)
- Pool and Hot Tub Alliance (PHTA), formerly The Association of Pool and Spa Professionals (APSP)

Code development committees evaluate and make recommendations regarding proposed changes to the codes. Their recommendations are then subject to public comment and council-wide votes. The ICC's governmental members—public safety officials who have no financial or business interest in the outcome—cast the final votes on proposed changes.

The I-Codes are subject to change through future code development cycles and by any governmental entity that enacts the code into law. For more information regarding the code development process, contact the Codes and Standards Development Department of the ICC at iccsafe.org/products-and-services/i-codes/code-development/.

While the I-Code development procedure is thorough and comprehensive, the ICC, its members and those participating in the development of the codes expressly disclaim any liability resulting from the publication or use of the I-Codes, or from compliance or noncompliance with their provisions. NO WARRANTY OF ANY KIND, IMPLIED, EXPRESSED OR STATUTORY, IS GIVEN WITH RESPECT TO THE I-CODES. The ICC does not have the power or authority to police or enforce compliance with the contents of the I-Codes.

Code Development Committee Responsibilities (Letter Designations in Front of Section Numbers)

In each cycle, proposed changes are considered by the Code Development Committee assigned to a specific code or subject matter. Committee Action Hearings result in recommendations regarding a proposal to the voting membership. Where changes to a code section are not considered by that code's own committee, the code section is preceded by a bracketed letter designation identifying a different committee. Bracketed letter designations for the I-Code committees are:

[A] = Administrative Code Development Committee
[BE] = IBC—Egress Code Development Committee
[BF] = IBC—Fire Safety Code Development Committee
[BG] = IBC—General Code Development Committee
[BS] = IBC—Structural Code Development Committee
[E] = Developed under the ICC's Standard Development Process
[EB] = International Existing Building Code Development Committee
[F] = International Fire Code Development Committee
[FG] = International Fuel Gas Code Development Committee
[M] = International Mechanical Code Development Committee
[P] = International Plumbing Code Development Committee
[SP] = International Swimming Pool and Spa Code Development Committee

For the development of the 2027 edition of the I-Codes, the ICC Board of Directors approved a standing motion from the Board Committee on the Long-Term Code Development Process to revise the code development cycle to incorporate two committee action hearings for each code group. This change expands the current process from two independent 1-year cycles to a single continuous 3-year cycle. There will be two groups of code development committees and they will meet in separate years. The current groups will be reworked. With the energy provisions of the *International Energy Conservation Code®* (IECC®) and Chapter 11 of the *International*

Residential Code® (IRC®) now moved to the Code Council's Standards Development Process, the reduced volume of code changes will be distributed between Groups A and B.

Code change proposals submitted for code sections that have a letter designation in front of them will be heard by the respective committee responsible for such code sections. Because different committees hold Committee Action Hearings in different years, proposals for most codes will be heard by committees in both the 2024 (Group A) and the 2025 (Group B) code development cycles. It is very important that anyone submitting code change proposals understands which code development committee is responsible for the section of the code that is the subject of the code change proposal.

Please visit the ICC website at iccsafe.org/products-and-services/i-codes/code-development/current-code-development-cycle for further information on the Code Development Committee responsibilities as it becomes available.

Coordination of the I-Codes

The coordination of technical provisions allows the I-Codes to be used as a complete set of complementary documents. Individual codes can also be used in subsets or as stand-alone documents. Some technical provisions that are relevant to more than one subject area are duplicated in multiple model codes.

Italicized Terms

Words and terms defined in Chapter 2, Definitions, are italicized where they appear in code text and the Chapter 2 definitions apply. Although care has been taken to ensure applicable terms are italicized, there may be instances where a defined term has not been italicized or where a term is italicized but the definition found in Chapter 2 is not applicable. For example, Chapter 2 of the *International Building Code*® (IBC®) contains a definition for "*Listed*" that is applicable to equipment, products and services. The term "listed" is also used in that code to refer to a list of items within the code or within a referenced document. For the latter, the Chapter 2 definition would not be applicable.

Adoption of International Code Council Codes and Standards

The International Code Council maintains a copyright in all of its codes and standards. Maintaining copyright allows the Code Council to fund its mission through sales of books in both print and digital formats. The Code Council welcomes incorporation by reference of its codes and standards by jurisdictions that recognize and acknowledge the Code Council's copyright in the codes and standards, and further acknowledge the substantial shared value of the public/private partnership for code development between jurisdictions and the Code Council. By making its codes and standards available for incorporation by reference, the Code Council does not waive its copyright in its codes and standards.

The Code Council's codes and standards may only be adopted by incorporation by reference in an ordinance passed by the governing body of the jurisdiction. "Incorporation by reference" means that in the adopting ordinance, the governing body cites only the title, edition, relevant sections or subsections (where applicable), and publishing information of the model code or standard, and the actual text of the model code or standard is not included in the ordinance (see graphic, "Adoption of International Code Council Codes and Standards"). The Code Council does not consent to the reproduction of the text of its codes or standards in any ordinance. If the governing body enacts any changes, only the text of those changes or amendments may be included in the ordinance.

ADOPTION OF INTERNATIONAL CODE COUNCIL CODES AND STANDARDS INCORPORATED BY REFERENCE

What does "incorporate by reference" mean? If a governmental agency or authority having jurisdiction (AHJ) over code adoption wishes to adopt a model code for legislative or regulatory purposes, it will enact an ordinance, regulation or law to incorporate by reference (IBR) the relevant code. The actual text of the model code is not included in the law, but the enacting law will include the full text of any changes or amendments enacted by the legislative body of the AHJ.

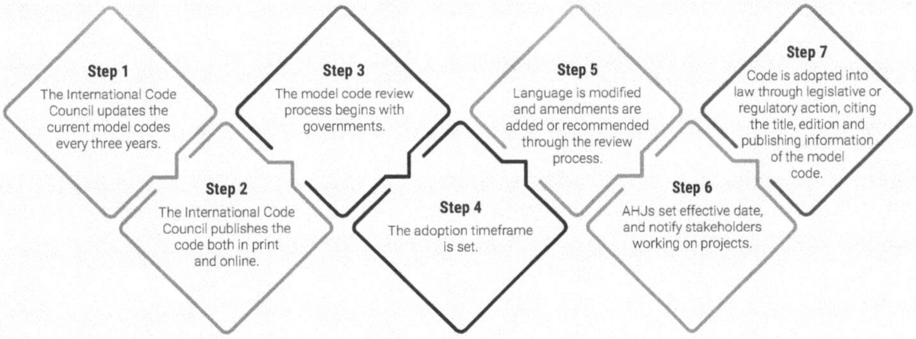

The Code Council also recognizes the need for jurisdictions to make laws accessible to the public. Accordingly, all I-Codes and I-Standards, along with the laws of many jurisdictions, are available to view for free at codes.iccsafe.org/codes/i-codes. These documents may also be purchased, in both digital and print versions, at shop.iccsafe.org.

To facilitate adoption, some I-Code sections contain blanks for fill-in information that needs to be supplied by the adopting jurisdiction as part of the adoption legislation. For example, the IWUIC contains:

Section 101.1. Insert: [NAME OF JURISDICTION]

Section 103.1. Insert: [NAME OF DEPARTMENT]

Section 109.3.7. Insert: [OFFENSE, DOLLAR AMOUNT, NUMBER OF DAYS]

For further information or assistance with adoption, including a sample ordinance, jurisdictions should contact the Code Council at incorporation@iccsafe.org.

For a list of frequently asked questions (FAQs) addressing a range of foundational topics about the adoption of model codes by jurisdictions and to learn more about the Code Council's code adoption resources, scan the QR code or visit iccsafe.org/code-adoption-resources.

INTRODUCTION TO THE INTERNATIONAL URBAN WILDLAND INTERFACE CODE

The IWUIC establishes minimum requirements for land use and the built environment in designated wildland-urban interface areas using prescriptive and performance-related provisions. It is founded on data collected from tests and fire incidents, technical reports and mitigation strategies from around the world.

Population growth and the expanding urban development into traditionally nonurban areas have increasingly brought humans into contact with wildfires. According to the National Fire Protection Agency (NFPA), every year, wildfires burn across the United States and a growing number of people are living where wildfires are a real risk. In 2018 more than 58,000 fires burned nearly 9 million acres across the US. More than 25,000 structures were destroyed, including 18,137 residences and 229 commercial structures. California accounted for the highest number of structures lost in one state due to the number of significant fires, including the Mendocino Complex, Carr, Camp and Woolsey fires.

The IWUIC is a model code that is intended to be adopted and used supplemental to the adopted building and fire codes of a jurisdiction. The unrestricted use of property in wildland-urban interface areas is a potential threat to life and property from fire and resulting erosion. The IWUIC has as its objective the establishment of minimum special regulations for the safeguarding of life and property from the intrusion of fire from wildland fire exposures and fire exposures from adjacent structures and to prevent structure fires from spreading to wildland fuels, even in the absence of fire department intervention.

Safeguards to prevent the occurrence of fires and to provide adequate fire protection facilities to control the spread of fire in wildland-urban interface areas are provided in a tiered manner commensurate with the relative level of hazard present.

ARRANGEMENT AND FORMAT OF THE 2024 IWUIC

The format of the IWUIC allows each chapter to be devoted to a particular subject. The following table shows how the IWUIC is divided. The subsequent table shows IWUIC requirements that are correlated with other I-Codes. The chapter synopses detail the scope and intent of the provisions of the IWUIC.

CHAPTER TOPICS	
CHAPTER	**SUBJECT**
1-2	Administration and Definitions
3-4	Wildland-Urban Interface Area Designation and Requirements
5	Building Construction Regulations
6	Fire Protection Requirements
7	Referenced Standards
Appendices A-I	Adoptable and Informational Appendices

Chapter 1 Scope and Administration.

Chapter 1 establishes the limits of applicability of the code and describes how the code is to be applied and enforced. The provisions of Chapter 1 establish the authority and duties of the code official appointed by the authority having jurisdiction and also establish the rights and privileges of the design professional, contractor and property owner.

Chapter 2 Definitions.

Chapter 2 is the repository of the definitions of terms used in the body of the code. The user of the code should be familiar with and consult this chapter because the definitions are essential to the correct interpretation of the code and because the user may not be aware that a term is defined.

Chapter 3 Wildland-Urban Interface Areas.

Chapter 3 provides for the fundamental aspect of applying the code—the legal declaration and establishment of wildland-urban interface areas within the adopting jurisdiction by the local legislative body. The provisions cover area analysis and declaration based on findings of fact (located in Appendix E), mapping of the area, legal recordation of the maps with the local keeper of records and the periodic review and reevaluation of the declared areas on a regular basis.

Chapter 4 Wildland-Urban Interface Area Requirements.

The requirements of Chapter 4 apply to all occupancies in the wildland-urban interface and pertain to all of the following:

1. Fire service access to the property that is to be protected, including fire apparatus access roads and off-road driveways.
2. Premises identification.
3. Key boxes to provide ready access to properties secured by gated roadways or other impediments to rapid fire service access.
4. Fire protection water supplies, including adequate water sources, pumper apparatus drafting sites, fire hydrant systems and system reliability.
5. Fire department access to equipment such as fire suppression equipment and fire hydrants.
6. Fire protection plans.

Chapter 5 Special Building Construction Regulations.

The regulations in Chapter 5 establish minimum standards for the location, design and construction of buildings and structures based on fire hazard severity in the wildland-urban interface.

The construction provisions of Chapter 5 are intended to supplement the requirements of the *International Building Code* and address mitigation of the unique hazards posed to buildings by wildfire and to reduce the hazards of building fires spreading to wildland fuels. This is accomplished by requiring ignition-resistant construction materials based on the hazard severity of the building site.

Chapter 6 Fire Protection Requirements.

Chapter 6 establishes minimum fire protection requirements to mitigate the hazards to life and property from fire in the wildland-urban interface. The chapter includes both design-oriented and prescriptive mitigation strategies to reduce the hazards of fire originating within a structure spreading to wildland and fire originating in wildland spreading to structures.

Chapter 7 Referenced Standards.

Chapter 7 lists all of the product and installation standards and codes that are referenced throughout Chapters 1 through 6 and includes identification of the promulgators and the section numbers in which the standards and codes are referenced. As stated in Section 102.4, these standards and codes become an enforceable part of the code (to the prescribed extent of the reference) as if printed in the body of the code.

Appendix A General Requirements.

Appendix A, while not part of the code, can become part of the code when specifically included in the adopting ordinance. Its purpose is to provide fire-protection measures supplemental to those found in Chapter 6 to reduce the threat of wildfire in a wildland-urban interface area and improve the capability for controlling such fires. This appendix includes detailed requirements for vegetation control; the code official's authority to close wildland-interface areas in times of high fire danger; control of fires, fireworks usage and other sources of ignition; storage of hazardous materials and combustibles; bans on the dumping of waste materials and ashes and coals in wildland-urban interface areas; protection of pumps and water supplies; and limits on temporary uses within the wildland-urban interface area.

Appendix B Vegetation Management Plan.

Appendix B, while not part of the code, can become part of the code when specifically included in the adopting ordinance. Its purpose is to provide criteria for submitting vegetation management plans, specifying their content and establishing a criterion for considering vegetation management as being a fuel modification.

Appendix C Fire Hazard Severity Form.

Appendix C, while not part of the code, can become part of the code (replacing Table 502.1) when specifically included in the adopting ordinance. Its purpose is to provide an alternative methodology to using Table 502.1 for analyzing the fire hazard severity of building sites using a preassigned value/scoring system for each feature that impacts the hazard level of a building site. Included in the evaluation are site access, types and management of vegetation, percentage of defensible space on the site, site topography, class of roofing and other construction materials used on the building (existing or to be constructed on the site), fire protection water supply, and whether utilities are installed above or below ground.

Appendix D Fire Danger Rating System.

Appendix D is an excerpt from the National Fire Danger Rating System (NFDRS), 1978, United States Department of Agriculture Forest Service, General Technical Report INT-39, and is for information purposes and is not intended for adoption. The fuel models that are included are only general descriptions because they represent all wildfire fuels from Florida to Alaska and from the East Coast to California.

NFDRS is a set of computer programs and algorithms that allow land management agencies to estimate today's or tomorrow's fire danger for a given rating area. The NFDRS characterizes fire danger by evaluating the approximate upper limit of fire behavior in a fire danger rating area during a 24-hour period based on fuels, topography and weather, or what is commonly called the fire triangle. Fire danger ratings are guides for initiating presuppression activities and selecting the appropriate level of initial response to a reported wildfire in lieu of detailed, site- and time-specific information.

Predicting the potential behavior and effects of wildland fire are essential tasks in fire management. Surface fire behavior and fire effects models and prediction systems are driven in part by fuelbed inputs such as load, bulk density, fuel particle size, heat content and moisture content. To facilitate use in models and systems, fuelbed inputs have been formulated into fuel models. A fuel model is a set of fuelbed inputs needed by a particular fire behavior or fire effects model. Different kinds of fuel models are used in fire spread models in a variety of fire behavior modeling systems. The fuel models in this appendix correlate with the light, medium and heavy fuel definitions found in Chapter 2.

Appendix E Findings of Fact.

Appendix E is an informational appendix that intends to provide a methodology for presenting the findings of fact that are required by Chapter 3 when a jurisdiction defines and establishes a wildland-urban interface area that will be the subject of regulation by the IWUIC. The development of a written "findings of fact" that justifies designation of wildland-interface areas by local jurisdictions requires that a certain amount of research and analysis be conducted to support a written finding that is both credible and professional. In the context of adopting a supplemental document such as the wildland-urban interface declaration, the writing of these findings is essential in creating the maps and overlap needed to use their specific options.

The purpose of this appendix is to provide an overview of how local officials could approach this process. There are three essential phenomena cited in some adoption statutes that vary from community to community: climate, topography and geography. Although it could be argued that other factors should be prioritized in statutes specific to any given locality, these three features are also consistent with standard code text that offers the opportunity to be more restrictive than local codes. The process demands a high level of professionalism to protect the jurisdiction's credibility in adopting more restrictive requirements. A superficial effort in preparing the findings of fact could jeopardize the proposed or adopted code restriction. Jurisdictions should devote a sufficient amount of time to draft the findings of fact to ensure that the facts are accurate, comprehensive and verifiable.

Appendix F Characteristics of Fire-Resistive Vegetation.

Appendix F is an informational appendix provided for the convenience of the code user. It is simply a compilation of the eight characteristics of fire-resistive vegetation that can be used effectively within wildland-urban interface areas to reduce the likelihood of fire spread through vegetation.

Appendix G Self-Defense Mechanism.

Appendix G is an informational appendix that provides discussion of some elements of the proposed self-defense mechanisms and their role in enhancing the protection of exposed structures in the wildland-urban interface. To accept alternative self-defense mechanisms, the code official must carefully examine whether these devices will be in place at the time of an event and whether they will assist or actually complicate the defense of the structure by fire suppression forces if they are available.

Appendix H International Wildland-Urban Interface Code Flowchart.

Appendix H is an informational appendix that is based on the "Decision Tree" concept and is intended to provide the code official with a graphical representation of how the IWUIC is to be applied in an orderly manner.

Appendix I Board of Appeals.

Appendix I contains the provisions for appeal and the establishment of a board of appeals. The provisions include the application for an appeal, the makeup of the board of appeals and the conduct of the appeal process.

RELOCATION OF TEXT OR TABLES

The following table indicates relocation of sections and tables in the 2024 edition of the IWUIC from the 2021 edition:

RELOCATIONS TABLE

2024 LOCATION	2021 LOCATION
102.5	104.4
102.6	104.5
102.7	102.5
102.8	102.6
104.2.1	105.2
104.2.2	105.3
104.2.3	105.1
104.3	104.6
104.4	110.3
104.8	104.3
104.10	104.7

CONTENTS

CHAPTER

1 SCOPE AND ADMINISTRATION

User notes:

About this chapter: *Chapter 1 establishes the limits of applicability of the code and describes how the code is to be applied and enforced. Chapter 1 is in two parts: Part 1–General Provisions (Sections 101–102) and Part 2—Administration and Enforcement (Sections 103–113). Section 101 identifies which buildings and structures come under its purview and references other International Codes as applicable.*

This code is intended to be adopted as a legally enforceable document and it cannot be effective without adequate provisions for its administration and enforcement. The provisions of Chapter 1 establish the authority and duties of the code official appointed by the authority having jurisdiction and also establish the rights and privileges of the design professional, contractor and property owner.

Section 105 was revised and relocated to Section 104 for the 2024 edition. For complete information, see the Relocations table in the Preface of this code.

QR code use: *A QR code is placed at the beginning of any section that has undergone technical revision. To see those revisions, scan the QR code with a smart device or enter the 7-digit code beneath the QR code at the end of the following URL: qr.iccsafe.org/ (see Formatting Changes to the 2024 International Codes for more information).*

PART 1—GENERAL PROVISIONS

SECTION 101—SCOPE AND GENERAL REQUIREMENTS

Scan for Changes 12f7d7a

[A] 101.1 Title. These regulations shall be known as the *Wildland-Urban Interface Code* of [NAME OF JURISDICTION], hereinafter referred to as "this code."

[A] 101.2 Scope. The provisions of this code shall apply to the construction, alteration, movement, repair, maintenance and use of any building, structure or premises within the *wildland-urban interface areas* in this jurisdiction.

Buildings or conditions in existence at the time of the adoption of this code are allowed to have their use or occupancy continued, if such condition, use or occupancy was legal at the time of the adoption of this code, provided that such continued use does not constitute a distinct danger to life or property.

Buildings or structures moved into or within the jurisdiction shall comply with the provisions of this code for new buildings or structures.

[A] 101.2.1 Appendices. Provisions in the appendices shall not apply unless specifically adopted.

[A] 101.3 Purpose. The purpose of this code is to establish minimum regulations for the safeguarding of life and for property protection. Regulations in this code are intended to mitigate the risk to life and structures from intrusion of fire from wildland fire exposures and fire exposures from adjacent structures and to mitigate structure fires from spreading to wildland fuels. The extent of this regulation is intended to be tiered commensurate with the relative level of hazard present.

The unrestricted use of property in *wildland-urban interface areas* is a potential threat to life and property from fire and resulting erosion. Safeguards to prevent the occurrence of fires and to provide adequate fire protection facilities to control the spread of fire in *wildland-urban interface areas* shall be in accordance with this code.

This code shall supplement the jurisdiction's building and fire codes, if such codes have been adopted, to provide for special regulations to mitigate the fire- and life-safety hazards of the *wildland-urban interface areas*.

[A] 101.4 Retroactivity. The provisions of the code shall apply to conditions arising after the adoption thereof, conditions not legally in existence at the adoption of this code and conditions that, in the opinion of the *code official*, constitute a distinct hazard to life or property.

Exception: Provisions of this code that specifically apply to existing conditions are retroactive.

[A] 101.5 Additions or alterations. Additions or alterations shall be permitted to be made to any building or structure without requiring the existing building or structure to comply with all of the requirements of this code, provided that the addition or alteration conforms to that required for a new building or structure.

Exception: Provisions of this code that specifically apply to existing conditions are retroactive.

Additions or alterations shall not be made to an existing building or structure that will cause the existing building or structure to be in violation of any of the provisions of this code nor shall such additions or alterations cause the existing building or structure to become unsafe. An unsafe condition shall be deemed to have been created if an addition or alteration will cause the existing building or structure to become structurally unsafe or overloaded; will not provide adequate access in compliance with the provisions of this code or will obstruct existing exits or access; will create a fire hazard; will reduce required fire resistance or will otherwise create conditions dangerous to human life.

[A] 101.6 Maintenance. Buildings, structures, landscape materials, vegetation, *defensible space* or other devices or safeguards required by this code shall be maintained in conformance to the code edition under which installed. The owner or the owner's authorized agent shall be responsible for the maintenance of buildings, structures, landscape materials and vegetation.

SECTION 102—APPLICABILITY

Scan for Changes 1c04970

[A] 102.1 General. Where there is a conflict between a general requirement and a specific requirement, the specific requirement shall govern. Where, in any specific case, different sections of this code specify different materials, methods of construction or other requirements, the most restrictive shall govern.

[A] 102.2 Other laws. The provisions of this code shall not be deemed to nullify any provisions of local, state or federal law.

[A] 102.3 Application of references. References to chapter or section numbers, or to provisions not specifically identified by number, shall be construed to refer to such chapter, section or provision of this code.

[A] 102.4 Referenced codes and standards. The codes and standards referenced in this code shall be those that are listed in Chapter 7 and such codes and standards shall be considered as part of the requirements of this code to the prescribed extent of each such reference and as further regulated in Sections 102.4.1 and 102.4.2.

[A] 102.4.1 Conflicts. Where conflicts occur between provisions of this code and the referenced standards, the provisions of this code shall govern.

[A] 102.4.2 Provisions in referenced codes and standards. Where the extent of the reference to a referenced code or standard includes subject matter that is within the scope of this code, the provisions of this code, as applicable, shall take precedence over the provisions in the referenced standard.

[A] 102.5 Subjects not regulated by this code. Where applicable standards or requirements are not set forth in this code, or are contained within other laws, codes, regulations, ordinances or policies adopted by the jurisdiction, compliance with applicable standards of other nationally recognized safety standards, as *approved*, shall be deemed as prima facie evidence of compliance with the intent of this code. Nothing herein shall derogate from the authority of the code official to determine compliance with codes or standards for those activities or installations within the code official's jurisdiction or responsibility.

[A] 102.6 Matters not provided for. Requirements that are essential for the public safety of an existing or proposed activity, building or structure, or for the safety of the occupants thereof, which are not specifically provided for by this code, shall be determined by the *code official* consistent with the necessity to establish the minimum requirements to safeguard the public health, safety and general welfare.

[A] 102.7 Partial invalidity. In the event that any part or provision of this code is held to be illegal or void, this shall not have the effect of making void or illegal any of the other parts or provisions.

[A] 102.8 Existing conditions. The legal occupancy or use of any structure or condition existing on the date of adoption of this code shall be permitted to continue without change, except as is specifically covered in this code, the *International Fire Code* or the *International Property Maintenance Code*, or as is deemed necessary by the *code official* for the general safety and welfare of the occupants and the public.

PART 2—ADMINISTRATION AND ENFORCEMENT

SECTION 103—CODE COMPLIANCE AGENCY

[A] 103.1 Creation of agency. The [INSERT NAME OF DEPARTMENT] is hereby created and the official in charge thereof shall be known as the *code official*. The function of the agency shall be the implementation, administration and enforcement of the provisions of this code.

[A] 103.2 Appointment. The *code official* shall be appointed by the chief appointing authority of the jurisdiction.

[A] 103.3 Deputies. In accordance with the prescribed procedures of this jurisdiction and with the concurrence of the appointing authority, the *code official* shall have the authority to appoint a deputy *code official*, other related technical officers, inspectors and other employees. Such employees shall have powers as delegated by the *code official*.

SECTION 104—DUTIES AND POWERS OF THE CODE OFFICIAL

Scan for Changes 1c245fe

[A] 104.1 Powers and duties of the code official. The *code official* is hereby authorized to enforce the provisions of this code.

[A] 104.2 Determination of compliance. The *code official* shall have the authority to determine compliance with this code, to render interpretations of this code and to adopt policies and procedures in order to clarify the application of its provisions. Such interpretations, policies and procedures:

1. Shall be in compliance with the intent and purpose of this code.
2. Shall not have the effect of waiving requirements specifically provided for in this code.

[A] 104.2.1 Technical assistance. To determine compliance with this code, the *code official* is authorized to require the owner, the owner's authorized agent or the person in possession or control of the building or premises to provide a technical opinion and report.

[A] 104.2.1.1 Costs. A technical opinion and report shall be provided without charge to the jurisdiction.

[A] 104.2.1.2 Preparer qualifications. The technical opinion and report shall be prepared by a qualified engineer, specialist, laboratory or fire safety specialty organization acceptable to the *code official.* The *code official* is authorized to require design submittals to be prepared by, and bear the stamp of, a *registered design professional*.

[A] 104.2.1.3 Content. The technical opinion and report shall analyze the properties of the design, operation or use of the building or premises, the facilities and appurtenances situated thereon and fuel management to identify and propose necessary recommendations.

[A] 104.2.1.4 Tests. Where there is insufficient evidence of compliance with the provisions of this code, the *code official* shall have the authority to require tests as evidence of compliance. Test methods shall be as specified in this code or by other recognized test standards. In the absence of recognized test standards, the *code official* shall approve the testing procedures. Such tests shall be performed by a party acceptable to the *code official*.

[A] 104.2.2 Alternative materials, design and methods. The provisions of this code are not intended to prevent the installation of any material or to prohibit any design or method of construction not specifically prescribed by this code, provided that any such alternative is not specifically prohibited by this code and has been *approved*.

[A] 104.2.2.1 Approval authority. An alternative material, design or method shall be approved where the *code official* finds that the proposed alternative is satisfactory and complies with Sections 104.2.2.2 through 104.2.2.7, as applicable.

[A] 104.2.2.2 Application and disposition. Where required, a request to use an alternative material, design or method of construction shall be submitted in writing to the *code official* for approval. Where the alternative material, design or method of construction is not approved, the *code official* shall respond in writing, stating the reasons the alternative was not approved.

[A] 104.2.2.3 Compliance with code intent. An alternative material, design or method of construction shall comply with the intent of the provisions of this code.

[A] 104.2.2.4 Equivalency criteria. An alternative material, design or method of construction shall, for the purpose intended, be not less than the equivalent of that prescribed in this code with respect to all of the following, as applicable:

1. Quality.
2. Strength.
3. Effectiveness.
4. Durability.
5. Safety, other than fire safety.
6. Fire safety.

[A] 104.2.2.5 Tests. Tests conducted to demonstrate equivalency in support of an alternative material, design or method of construction application shall be of a scale that is sufficient to predict performance of the end use configuration. Tests shall be performed by a party acceptable to the *code official*.

[A] 104.2.2.5.1 Fire tests. Tests conducted to demonstrate equivalent fire safety in support of an alternative material, design or method of construction application shall be of a scale that is sufficient to predict fire safety performance of the end use configuration. Tests shall be performed by a party acceptable to the code official.

[A] 104.2.2.6 Reports. Supporting data, where necessary to assist in the approval of materials or assemblies not specifically provided for in this code, shall comply with Sections 104.2.2.6.1 and 104.2.2.6.2.

[A] 104.2.2.6.1 Evaluation reports. Evaluation reports shall be issued by an *approved agency* and use of the evaluation report shall require approval by the code official for the installation. The alternate material, design or method of construction and product evaluated shall be within the scope of the code official's recognition of the *approved agency*. Criteria used for the evaluation shall be identified within the report and, where required, provided to the code official.

[A] 104.2.2.6.2 Other reports. Reports not complying with Section 104.2.2.6.1 shall describe criteria, including but not limited to any referenced testing or analysis, used to determine compliance with code intent and justify code equivalence. The report shall be prepared by a qualified engineer, specialist, laboratory or fire safety specialty organization acceptable to the code official. The *code official* is authorized to require design submittals to be prepared by, and bear the stamp of, a *registered design professional*.

[A] 104.2.2.7 Peer review. The *code official* is authorized to require submittal of a peer review report in conjunction with a request to use an alternative material, design or method of construction, prepared by a peer reviewer that is approved by the *code official*.

[A] 104.2.3 Modifications. Where there are practical difficulties involved in carrying out the provisions of this code, the *code official* shall have the authority to grant modifications for individual cases, provided that the *code official* shall first find that one or more special individual reasons make the strict letter of this code impractical, that the modification is in conformance with the intent and purpose of this code, and that such modification does not lessen health, life and fire safety requirements. The details of the written request and action granting modifications shall be recorded and entered into the files of the code enforcement agency.

[A] 104.3 Applications and permits. The *code official* is authorized to receive applications, review construction documents and issue permits for construction regulated by this code, issue permits for operations regulated by this code, inspect the premises for which such permits have been issued and enforce compliance with the provisions of this code.

[A] 104.4 Right of entry. Where it is necessary to make an inspection to enforce the provisions of this code, or where the *code official* has reasonable cause to believe that there exists in a structure or on any premises a condition that is contrary to or in violation of this code that makes the structure or premises unsafe, dangerous or hazardous, the *code official* is authorized to enter the structure or premises at all reasonable times to inspect or to perform the duties imposed by this code. If such structure or premises is occupied, the *code official* shall present proper credentials to the occupant and request entry. If such structure or premises is unoccupied, the *code official* shall first make a reasonable effort to locate the owner, the owner's authorized agent or other person having charge or control of the structure or premises and request entry. If such entry is refused, then the *code official* shall have recourse to every remedy provided by law to secure entry.

[A] 104.4.1 Warrant. Where the code official has first obtained a proper inspection warrant or other remedy provided by law to secure entry, an owner, the owner's authorized agent, occupant or person having charge, care or control of the structure or premises shall not fail or neglect, after a proper request is made as herein provided, to permit entry therein by the code official for the purposes of inspection and examination pursuant to this code.

[A] 104.5 Identification. The *code official* shall carry proper identification when inspecting structures or premises in the performance of duties under this code.

[A] 104.6 Notices and orders. The code official shall issue all necessary notices or orders to ensure compliance with this code. Notices of violations shall be in accordance with Section 110.2.

[A] 104.7 Official records. The *code official* shall keep official records as required by Sections 104.7.1 through 104.7.5. Such official records shall be retained for not less than 5 years or for as long as the structure or activity to which such records relate remains in existence, unless otherwise provided by other regulations.

[A] 104.7.1 Approvals. A record of approvals shall be maintained by the *code official* and shall be available for public inspection during business hours in accordance with applicable laws.

[A] 104.7.2 Inspections. The *code official* shall keep a record of each inspection made, including notices and orders issued, showing the findings and disposition of each.

[A] 104.7.3 Code alternatives and modifications. Application for alternative materials, design and methods of construction and equipment in accordance with Section 104.2.2; modifications in accordance with Section 104.2.3; and documentation of the final decision of the *code official* for either shall be in writing and shall be retained in the official records.

[A] 104.7.4 Tests. The *code official* shall keep a record of tests conducted to comply with Sections 104.2.1.4 and 104.2.2.5.

[A] 104.7.5 Fees. The *code official* shall keep a record of fees collected and refunded in accordance with Section 108.

[A] 104.8 Liability. The *code official*, member of the board of appeals or employee charged with the enforcement of this code, while acting for the jurisdiction, in good faith and without malice in the discharge of the duties required by this code or other pertinent law or ordinance, shall not thereby be rendered personally liable, either civilly or criminally, and is hereby relieved from all personal liability for any damage accruing to persons or property as a result of an act or by reason of any act or omission in the discharge of official duties.

[A] 104.8.1 Legal defense. Any suit or criminal complaint instituted against any officer or employee because of an act performed by that officer or employee in the lawful discharge of duties and under the provisions of this code or other laws or ordinances implemented through the enforcement of this code shall be defended by legal representatives of the jurisdiction until final termination of the proceedings. The code official or any subordinate shall not be liable for costs in an action, suit or proceeding that is instituted in pursuance of the provisions of this code.

[A] 104.9 Approved materials and equipment. Materials, equipment and devices approved by the code official shall be constructed and installed in accordance with such approval.

[A] 104.9.1 Materials and equipment reuse. Materials, equipment and devices shall not be reused unless such elements are in good working order and *approved*.

[A] 104.10 Other agencies. When requested to do so by the *code official*, other officials of this jurisdiction shall assist and cooperate with the *code official* in the discharge of the duties required by this code.

SECTION 105—PERMITS

[A] 105.1 General. Where not otherwise provided in the requirements of the *International Building Code* or *International Fire Code*, permits are required in accordance with Sections 105.2 through 105.10.

[A] 105.2 Permits required. Unless otherwise exempted, buildings or structures regulated by this code shall not be erected, constructed, altered, repaired, moved, removed, converted, demolished or changed in use or occupancy unless a separate permit for each building or structure has first been obtained from the *code official*.

For buildings or structures erected for temporary uses, see Section A108.3.

Where required by the *code official*, a permit shall be obtained for the following activities, operations, practices or functions within a *wildland-urban interface area*:

1. Automobile wrecking yard.
2. Candles and open flames in assembly areas.
3. Explosives or blasting agents.

4. Fireworks.
5. Flammable or combustible liquids.
6. Hazardous materials.
7. Liquefied petroleum gases.
8. Lumberyards.
9. Motor vehicle fuel-dispensing stations.
10. Open burning.
11. Pyrotechnical special effects material.
12. Tents, canopies and temporary membrane structures.
13. Tire storage.
14. Welding and cutting operations.

[A] 105.3 Work exempt from permit. Unless otherwise provided in the requirements of the *International Building Code* or *International Fire Code*, a permit shall not be required for the following:

1. One-story detached accessory structures used as tool and storage sheds, playhouses and similar uses, provided that the floor area does not exceed 120 square feet (11.15 m^2) and the structure is located more than 50 feet (15 240 mm) from the nearest adjacent structure.
2. Fences not over 6 feet (1829 mm) high.

Exemption from the permit requirements of this code shall not be deemed to grant authorization for any work to be done in any manner in violation of the provisions of this code or any other laws or ordinances of this jurisdiction.

The *code official* is authorized to stipulate conditions for permits. Permits shall not be issued where public safety would be at risk, as determined by the *code official*.

[A] 105.4 Permit application. To obtain a permit, the applicant shall first file an application therefor in writing on a form furnished by the code enforcement agency for that purpose. Every such application shall:

1. Identify and describe the work, activity, operation, practice or function to be covered by the permit for which application is made.
2. Describe the land on which the proposed work, activity, operation, practice or function is to be done by legal description, street address or similar description that will readily identify and definitely locate the proposed building, work, activity, operation, practice or function.
3. Indicate the use or occupancy for which the proposed work, activity, operation, practice or function is intended.
4. Be accompanied by plans, diagrams, computation and specifications and other data as required in Section 106 of this code.
5. State the valuation of any new building or structure or any addition, remodeling or alteration to an existing building.
6. Be signed by the applicant or the applicant's authorized agent.
7. Give such other data and information as required by the *code official*.

[A] 105.4.1 Preliminary inspection. Before a permit is issued, the *code official* is authorized to inspect and approve the systems, equipment, buildings, devices, premises and spaces or areas to be used.

[A] 105.4.2 Time limitation of application. An application for a permit for any proposed work shall be deemed to have been abandoned 180 days after the date of filing, unless such application has been pursued in good faith or a permit has been issued; except that the *code official* is authorized to grant one or more extensions of time for additional periods not exceeding 180 days each. The extension shall be requested in writing and justifiable cause demonstrated.

[A] 105.5 Permit approval. Before a permit is issued, the *code official*, or an authorized representative, shall review and approve permitted uses, occupancies or structures. Where laws or regulations are enforceable by other agencies or departments, a joint approval shall be obtained from agencies or departments concerned.

[A] 105.6 Permit issuance. The application, plans, specifications and other data filed by an applicant for a permit shall be reviewed by the *code official*. If the *code official* finds that the work described in an application for a permit and the plan, specifications and other data filed therewith conform to the requirements of this code, the *code official* is allowed to issue a permit to the applicant.

When the *code official* issues the permit, the *code official* shall endorse in writing or stamp the plans and specifications APPROVED. Such *approved* plans and specifications shall not be changed, modified or altered without authorization from the *code official*, and work regulated by this code shall be done in accordance with the *approved* plans.

[A] 105.6.1 Refusal to issue a permit. Where the application or construction documents do not conform to the requirements of pertinent laws, the *code official* shall reject such application in writing, stating the reasons therefor.

[A] 105.7 Validity of permit. The issuance or granting of a permit or approval of plans, specifications and computations shall not be construed to be a permit for, or an approval of, any violation of any of the provisions of this code or of any other ordinance of the jurisdiction. Permits presuming to give authority to violate or conceal the provisions of this code or other ordinances of the jurisdiction shall not be valid.

[A] 105.8 Expiration. Every permit issued by the *code official* under the provisions of this code shall expire by limitation and become null and void if the building, use or work authorized by such permit is not commenced within 180 days from the date of such permit, or if the building, use or work authorized by such permit is suspended or abandoned at any time after the work is commenced for a period of 180 days.

Any permittee holding an unexpired permit is allowed to apply for an extension of the time within which work is allowed to commence under that permit where the permittee is unable to commence work within the time required by this section for good and satisfactory reasons. The *code official* is authorized to extend the time for action by the permittee for a period not exceeding 180 days on written request by the permittee showing that circumstances beyond the control of the permittee have prevented action from being taken. Permits shall not be extended more than once.

[A] 105.9 Retention of permits. Permits shall at all times be kept on the premises designated therein and shall at all times be subject to inspection by the *code official* or other authorized representative.

[A] 105.10 Revocation of permits. Permits issued under this code can be suspended or revoked where it is determined by the *code official* that:

1. It is used by a person other than the person to whom the permit was issued.
2. It is used for a location other than that for which the permit was issued.
3. Any of the conditions or limitations set forth in the permit have been violated.
4. The permittee fails, refuses or neglects to comply with any order or notice duly served on him or her under the provisions of this code within the time provided therein.
5. There has been any false statement or misrepresentation as to material fact in the application or plans on which the permit or application was made.
6. The permit is issued in error or in violation of any other ordinance, regulations or provisions of this code.

The *code official* is allowed to, in writing, suspend or revoke a permit issued under the provisions of this code whenever the permit is issued in error or on the basis of incorrect information supplied, or in violation of any ordinance or regulation or any of the provisions of this code.

SECTION 106—CONSTRUCTION DOCUMENTS

[A] 106.1 General. Plans, engineering calculations, diagrams and other data shall be submitted in not fewer than two sets, or in a digital format where allowed by the building official, with each application for a permit. The construction documents shall be prepared by a registered design professional where required by the statutes of the jurisdiction in which the project is to be constructed. Where special conditions exist, the *code official* is authorized to require additional documents to be prepared by a registered design professional.

Exception: Submission of plans, calculations, construction inspection requirements and other data, if it is found that the nature of the work applied for is such that reviewing of plans is not necessary to obtain compliance with this code.

[A] 106.2 Information on plans and specifications. Plans and specifications shall be drawn to scale on substantial paper or cloth and shall be of sufficient clarity to indicate the location, nature and extent of the work proposed, and show in detail that it will conform to the provisions of this code and relevant laws, ordinances, rules and regulations.

[A] 106.3 Site plan. In addition to the requirements for plans in the *International Building Code*, site plans shall include topography, width and percent of grade of access roads, landscape and vegetation details, locations of structures or building envelopes, existing or proposed overhead utilities, occupancy classification of buildings, types of ignition-resistant construction of buildings, structures and their appendages, roof classification of buildings and site water supply systems. The *code official* is authorized to waive or modify the requirement for a site plan where the application for permit is for alteration or repair or where otherwise warranted.

[A] 106.4 Vegetation management plans. Where utilized by the permit applicant pursuant to Section 502, vegetation management plans shall be prepared and shall be submitted to the *code official* for review and approval as part of the plans required for a permit.

[A] 106.5 Fire protection plan. Where required by the *code official* pursuant to Section 405, a *fire protection plan* shall be prepared and shall be submitted to the *code official* for review and *approved* as a part of the plans required for a permit.

[A] 106.6 Other data and substantiation. Where required by the *code official*, the plans and specifications shall include classification of fuel loading, fuel model light, medium or heavy, and substantiating data to verify classification of fire-resistive vegetation.

[A] 106.7 Vicinity plan. In addition to the requirements for site plans, plans shall include details regarding the vicinity within 300 feet (91 440 mm) of lot lines, including other structures, slope, vegetation, *fuel breaks*, water supply systems and access roads.

[A] 106.8 Retention of plans. One set of *approved* plans, specifications and computations shall be retained by the *code official* for a period of not less than 180 days from date of completion of the permitted work or as required by state or local laws; and one set of *approved* plans and specifications shall be returned to the applicant, and said set shall be kept on the site of the building, use or work at all times during which the work authorized thereby is in progress.

[A] 106.9 Examination of documents. The *code official* shall examine or cause to be examined the accompanying construction documents and shall ascertain by such examinations whether the construction indicated and described is in accordance with the requirements of this code and other pertinent laws or ordinances.

[A] 106.10 Amended construction documents. Work shall be installed in accordance with the *approved* construction documents, and changes made during construction that are not in compliance with the *approved* documents shall be resubmitted for *approval* as an amended set of construction documents.

[A] 106.11 Previous approvals. This code shall not require changes in the construction documents, construction or designated occupancy of a structure for which a lawful permit has been heretofore issued or otherwise lawfully authorized, and the construction of which has been pursued in good faith within 180 days after the effective date of this code and has not been abandoned.

[A] 106.12 Phased approval. The *code official* is authorized to issue a permit for the construction of foundations or any other part of a building or structure before the construction documents for the whole building or structure have been submitted, provided that adequate information and detailed statements have been filed complying with pertinent requirements of this code. The holder of such permit for the foundation or other parts of a building or structure shall proceed at the holder's own risk with the building operation and without assurance that a permit for the entire structure will be granted.

SECTION 107—TEMPORARY USES, EQUIPMENT AND SYSTEMS

Scan for Changes 67983d5

[A] 107.1 General. The *code official* is authorized to issue a permit for temporary uses, equipment and systems. Such permits shall be limited as to time of service, but shall not be permitted for more than 180 days. The *code official* is authorized to grant extensions for demonstrated cause.

[A] 107.2 Conformance. Temporary uses, equipment and systems shall conform to the requirements of this code as necessary to ensure health, safety and general welfare.

[A] 107.3 Temporary service utilities. The *code official* is authorized to give permission to temporarily supply service utilities in accordance with Section 111.

[A] 107.4 Termination of approval. The *code official* is authorized to terminate such permit for temporary uses, equipment and systems and to order the same to be discontinued.

SECTION 108—FEES

Scan for Changes 2111368

[A] 108.1 Payment of fees. A permit shall not be valid until the fees prescribed by law have been paid. An amendment to a permit shall not be released until the additional fee, if any, has been paid

[A] 108.2 Schedule of permit fees. Where a permit is required, a fee for each permit shall be paid as required, in accordance with the schedule as established by the applicable governing authority.

[A] 108.3 Permit valuations. The applicant for a permit shall provide an estimated value of the work for which the permit is being issued at time of application. Such estimated valuations shall include the total value of work, including materials and labor, for which the permit is being issued. Where, in the opinion of the applicable governing authority, the valuation is underestimated, the permit shall be denied, unless the applicant can show detailed estimates acceptable to the applicable governing authority. The applicable governing authority shall have the authority to adjust the final valuation for permit fees.

[A] 108.4 Work commencing before permit issuance. Any person who commences any work before obtaining the necessary permits shall be subject to a fee established by the applicable governing authority, which shall be in addition to the required permit fees.

[A] 108.5 Related fees. The payment of the fee for the construction, alteration, removal or demolition of work done in connection to or concurrently with the work or activity authorized by a permit shall not relieve the applicant or holder of the permit from the payment of other fees that are prescribed by law.

[A] 108.6 Refunds. The applicable governing authority is authorized to establish a refund policy.

SECTION 109—INSPECTION AND ENFORCEMENT

[A] 109.1 Inspection. Inspections shall be in accordance with Sections 109.1.1 through 109.1.4.3.

[A] 109.1.1 General. Construction or work for which a permit is required by this code shall be subject to inspection by the *code official* and such construction or work shall remain visible and able to be accessed for inspection purposes until *approved* by the *code official*.

It shall be the duty of the permit applicant to cause the work to remain visible and able to be accessed for inspection purposes. Neither the *code official* nor the jurisdiction shall be liable for expense entailed in the removal or replacement of any material required to allow inspection.

Approval as a result of an inspection shall not be construed to be an approval of a violation of the provisions of this code or of other ordinances of the jurisdiction. Inspections presuming to give authority to violate or cancel the provisions of this code or of other ordinances of the jurisdiction shall not be valid.

Where required by the *code official*, a survey of the lot shall be provided to verify that the mitigation features are provided and the building or structure is located in accordance with the *approved* plans.

[A] 109.1.2 Authority to inspect. The *code official* shall inspect, as often as necessary, buildings and premises, including such other hazards or appliances designated by the *code official* for the purpose of ascertaining and causing to be corrected any conditions that could reasonably be expected to cause fire or contribute to its spread, or any violation of the purpose of this code and of any other law or standard affecting fire safety.

[A] 109.1.2.1 Approved inspection agencies. The *code official* is authorized to accept reports of *approved* inspection agencies, provided that such agencies satisfy the requirements as to qualifications and reliability.

[A] 109.1.2.2 Inspection requests. It shall be the duty of the holder of the permit or their duly authorized agent to notify the *code official* when work is ready for inspection. It shall be the duty of the permit holder to provide access to and means for inspections of such work that are required by this code.

[A] 109.1.2.3 Approval required. Work shall not be done beyond the point indicated in each successive inspection without first obtaining the approval of the *code official*. The *code official*, upon notification, shall make the requested inspections and shall either indicate the portion of the construction that is satisfactory as completed, or notify the permit holder or his or her agent wherein the same fails to comply with this code. Any portions that do not comply shall be corrected and such portion shall not be covered or concealed until authorized by the *code official*.

[A] 109.1.3 Reinspections. To determine compliance with this code, the *code official* can cause a structure to be reinspected. A fee can be assessed for each inspection or reinspection where work for which inspection is called is not complete or where corrections called for are not made.

Reinspection fees can be assessed where the *approved* plans are not readily available to the inspector, for failure to provide access on the date for which inspection is requested or for deviating from plans requiring the approval of the *code official*.

To obtain a reinspection, the applicant shall pay the reinspection fee as set forth in the fee schedule adopted by the jurisdiction. Where reinspection fees have been assessed, additional inspection of the work will not be performed until the required fees have been paid.

[A] 109.1.4 Testing. Installations shall be tested as required in this code and in accordance with Sections 109.1.4.1 through 109.1.4.3. Tests shall be made by the permit holder or authorized agent and observed by the *code official*.

[A] 109.1.4.1 New, altered, extended or repaired installations. New installations and parts of existing installations that have been altered, extended, renovated or repaired, shall be tested as prescribed herein to disclose defects.

[A] 109.1.4.2 Apparatus, instruments, material and labor for tests. Apparatus, instruments, material and labor required for testing an installation or part thereof shall be furnished by the permit holder or authorized agent.

[A] 109.1.4.3 Reinspection and testing. Where any work or installation does not pass an initial test or inspection, the necessary corrections shall be made so as to achieve compliance with this code. The work or installation shall then be resubmitted to the *code official* for inspection and testing.

[A] 109.2 Enforcement. Enforcement shall be in accordance with Sections 109.2.1 and 109.2.2.

[A] 109.2.1 Authorization to issue corrective orders and notices. Where the *code official* finds any building or premises that are in violation of this code, the *code official* is authorized to issue corrective orders and notices.

[A] 109.2.2 Service of orders and notices. Orders and notices authorized or required by this code shall be given or served on the owner, the owner's authorized agent, operator, occupant or other person responsible for the condition or violation either by verbal notification, personal service, or delivering the same to, and leaving it with, a person of suitable age and discretion on the premises; or, if such person is not found on the premises, by affixing a copy thereof in a conspicuous place on the door to the entrance of said premises and by mailing a copy thereof to such person by registered or certified mail to the person's last known address.

Orders or notices that are given verbally shall be confirmed by service in writing as herein provided.

[A] 109.3 Compliance with orders and notices. Compliance with orders and notices shall be in accordance with Sections 109.3.1 through 109.3.8.

[A] 109.3.1 General compliance. Orders and notices issued or served as provided by this code shall be complied with by the owner, the owner's authorized agent, operator, occupant or other person responsible for the condition or violation to which the corrective order or notice pertains.

If the building or premises is not occupied, then such corrective orders or notices shall be complied with by the owner or the owner's authorized agent.

[A] 109.3.2 Compliance with tags. A building or premises shall not be used when in violation of this code as noted on a tag affixed in accordance with Section 109.3.1.

[A] 109.3.3 Removal and destruction of signs and tags. A sign or tag posted or affixed by the *code official* shall not be mutilated, destroyed or removed without authorization by the *code official*.

[A] 109.3.4 Citations. Persons operating or maintaining an occupancy, premises or vehicle subject to this code who allow a hazard to exist or fail to take immediate action to abate a hazard on such occupancy, premises or vehicle when ordered or notified to do so by the *code official* shall be guilty of a misdemeanor.

[A] 109.3.5 Unsafe conditions. Buildings, structures or premises that constitute a fire hazard or are otherwise dangerous to human life, or that in relation to existing use constitute a hazard to safety or health or public welfare, by reason of inadequate maintenance, dilapidation, obsolescence, fire hazard, disaster damage or abandonment as specified in this code or any other ordinance, are unsafe conditions. Unsafe buildings or structures shall not be used. Unsafe buildings are hereby declared to be public nuisances and shall be abated by repair, rehabilitation, demolition or removal, pursuant to applicable state and local laws and codes.

[A] 109.3.5.1 Record. The *code official* shall cause a report to be filed on an unsafe condition. The report shall state the occupancy of the structure and the nature of the unsafe condition.

[A] 109.3.5.2 Notice. Where an unsafe condition is found, the *code official* shall serve on the owner, owner's authorized agent or person in control of the building, structure or premises, a written notice that describes the condition deemed unsafe and specifies the required repairs or improvements to be made to abate the unsafe condition, or requires the unsafe structure to be demolished. Such notice shall require the person thus notified, or their designee, to declare to the *code official* within a stipulated time, acceptance or rejection of the terms of the order.

[A] 109.3.5.2.1 Method of service. Such notice shall be deemed properly served where a copy thereof is served by one of the following methods:

1. Delivered to the owner or the owner's authorized agent personally.
2. Sent by certified or registered mail addressed to the owner or the owner's authorized agent at the last known address with a return receipt requested.
3. Delivered in any other manner as prescribed by local law.

Where the certified or registered letter is returned showing that the letter was not delivered, a copy thereof shall be posted in a conspicuous place in or about the structure affected by such notice. Service of such notice in the foregoing manner on the owner's authorized agent or on the person responsible for the structure shall constitute service of notice on the owner.

[A] 109.3.5.3 Placarding. Upon failure of the owner, the owner's authorized agent or the person responsible to comply with the notice provisions within the time given, the *code official* shall post on the premises or on defective equipment a placard bearing the word "UNSAFE" and a statement of the penalties provided for occupying the premises, operating the equipment or removing the placard.

[A] 109.3.5.3.1 Placard removal. The *code official* shall remove the unsafe condition placard whenever the defect or defects on which the unsafe condition and placarding action were based have been eliminated. Any person who defaces or removes an unsafe condition placard without the approval of the *code official* shall be subject to the penalties provided by this code.

[A] 109.3.5.4 Abatement. The owner, the owner's authorized agent, operator or occupant of a building, structure or premises deemed unsafe by the *code official* shall abate, correct or cause to be abated or corrected such unsafe conditions either by repair, rehabilitation, demolition or other *approved* corrective action.

[A] 109.3.5.5 Summary abatement. Where conditions exist that are deemed hazardous to life and property, the *code official* is authorized to abate or correct summarily such hazardous conditions that are in violation of this code.

[A] 109.3.5.6 Evacuation. The *code official* shall be authorized to order the immediate evacuation of any occupied building, structure or premises deemed unsafe where such hazardous conditions exist that present imminent danger to the occupants. Persons so notified shall immediately leave the structure or premises and shall not enter or reenter until authorized to do so by the *code official*.

[A] 109.3.6 Prosecution of violation. If the notice of violation is not complied with promptly, the *code official* is authorized to request the legal counsel of the jurisdiction to institute the appropriate proceeding at law or in equity to restrain, correct or abate such violation, or to require the removal or termination of the unlawful occupancy of the building or structure in violation of the provisions of this code or of the order or direction made pursuant thereto.

[A] 109.3.7 Violation penalties. Persons who shall violate a provision of this code or shall fail to comply with any of the requirements thereof or who shall erect, install, alter, repair or do work in violation of the *approved* construction documents or directive of the *code official*, or of a permit or certificate used under provisions of this code, shall be guilty of a **[SPECIFY OFFENSE]**, punishable by a fine of not more than **[AMOUNT]** dollars or by imprisonment not exceeding **[NUMBER OF DAYS]**, or both such fine and imprisonment. Each day that a violation continues after due notice has been served shall be deemed a separate offense.

[A] 109.3.8 Abatement of violation. In addition to the imposition of the penalties herein described, the *code official* is authorized to institute appropriate action to prevent unlawful construction or to restrain, correct or abate a violation; or to prevent illegal occupancy of a structure or premises; or to stop an illegal act, conduct of business or occupancy of a structure on or about any premises.

SECTION 110—CERTIFICATE OF COMPLETION

[A] 110.1 General. A building, structure or premises shall not be used or occupied, and a change in the existing use or occupancy classification of a building, structure, premise or portion thereof shall not be made until the *code official* has issued a certificate of completion therefor as provided herein. The certificate of occupancy shall not be issued until the certificate of completion indicating that the project is in compliance with this code has been issued by the *code official*.

[A] 110.2 Certificate of occupancy. Issuance of a certificate of occupancy shall not be construed as an approval of a violation of the provisions of this code or of other pertinent laws and ordinances of the jurisdiction. Certificates presuming to give authority to violate or cancel the provisions of this code or other laws or ordinances of the jurisdiction shall not be valid.

Exceptions:

1. Certificates of occupancy are not required for work exempt from permits under Section 105.3.
2. Accessory structures.

[A] 110.3 Temporary occupancy. The *code official* is authorized to issue a temporary certificate of occupancy before the completion of the entire work covered by the permit, provided that such portion or portions shall be occupied safely. The *code official* shall set a time period during which the temporary certificate of occupancy is valid.

[A] 110.4 Revocation. The *code official* is authorized to, in writing, suspend or revoke a certificate of occupancy or completion issued under the provisions of this code wherever the certificate is issued in error, on the basis of incorrect information supplied, or where it is determined that the building or structure, premise or portion thereof is in violation of any ordinance or regulation or any of the provisions of this code.

SECTION 111—SERVICE UTILITIES

[A] 111.1 Connection of service utilities. A person shall not make connections from a utility, source of energy, fuel, power, water system or sewer system to any building or system that is regulated by this code for which a permit is required until authorized by the *code official*.

[A] 111.2 Temporary connection. The *code official* shall have the authority to authorize the temporary connection of the building or system to the utility, source of energy, fuel, power, water system or sewer system for the purpose of testing systems or for use under a temporary approval.

[A] 111.3 Authority to disconnect service utilities. The *code official* shall have the authority to authorize disconnection of utility service to the building, structure or system regulated by this code and the referenced codes and standards in case of emergency where necessary to eliminate an immediate hazard to life or property or where such utility connection has been made without the approval required by Sections 111.1 and 111.2. The *code official* shall notify the serving utility and, where possible, the owner or the owner's authorized agent and the occupant of the building, structure or service system of the decision to disconnect prior to taking such action. If not notified prior to disconnection, the owner, the owner's authorized agent or the occupant of the building, structure or service system shall be notified in writing as soon as practical thereafter.

SECTION 112—MEANS OF APPEALS

[A] 112.1 General. In order to hear and decide appeals of orders, decisions or determinations made by the *code official* relative to the application and interpretation of this code, there shall be and is hereby created a board of appeals. The board of appeals shall be appointed by the applicable governing authority and shall hold office at its pleasure. The board shall adopt rules of procedure for conducting its business and shall render all decisions and findings in writing to the appellant, with a duplicate copy to the *code official*.

Scan for Changes
d50c335

[A] 112.2 Limitations on authority. An application for appeal shall be based on a claim that the true intent of this code or the rules legally adopted thereunder have been incorrectly interpreted, the provisions of this code do not fully apply or an equivalent or better form of construction is proposed. The board shall not have authority to waive requirements of this code.

[A] 112.3 Qualifications. The board of appeals shall consist of members who are qualified by experience and training on matters pertaining to the provisions of this code and are not employees of the jurisdiction.

[A] 112.4 Administration. The *code official* shall take action without delay in accordance with the decision of the board.

SECTION 113—STOP WORK ORDER

[A] 113.1 Authority. Where the *code official* finds any work regulated by this code being performed in a manner contrary to the provisions of this code or in a dangerous or unsafe manner, the *code official* is authorized to issue a stop work order.

[A] 113.2 Issuance. The stop work order shall be in writing and shall be given to the owner of the property, the owner's authorized agent or the person performing the work. Upon issuance of a stop work order, the cited work shall immediately cease. The stop work order shall state the reason for the order and the conditions under which the cited work is authorized to resume.

[A] 113.3 Emergencies. Where an emergency exists, the *code official* shall not be required to give a written notice prior to stopping the work.

[A] 113.4 Failure to comply. Any person who shall continue any work after having been served with a stop work order, except such work as that person is directed to perform to remove a violation or unsafe condition, shall be subject to fines established by the authority having jurisdiction.

CHAPTER

2 DEFINITIONS

User notes:

About this chapter: *Codes, by their very nature, are technical documents. Every word, term and punctuation mark can add to or change the meaning of a technical requirement. It is necessary to maintain a consensus on the specific meaning of each term contained in the code. Chapter 2 performs this function by stating clearly what specific terms mean for the purpose of the code.*

SECTION 201—GENERAL

201.1 Scope. Unless otherwise expressly stated, the following words and terms shall, for the purposes of this code, have the meanings shown in this chapter.

201.2 Interchangeability. Words stated in the present tense include the future; words stated in the masculine gender include the feminine and neuter; and the singular number includes the plural and the plural the singular.

201.3 Terms defined in other codes. Where terms are not defined in this code and are defined in other International Codes, such terms shall have the meanings ascribed to them as in those codes.

201.4 Terms not defined. Where terms are not defined through the methods authorized by this section, such terms shall have their ordinarily accepted meanings such as the context implies.

SECTION 202—DEFINITIONS

Scan for Changes

b826d47

ACCESSORY STRUCTURE. A building or structure used to shelter or support any material, equipment, chattel or occupancy other than a habitable building.

[A] APPROVED. Acceptable to the *code official*.

[A] APPROVED AGENCY. An established and recognized organization that is regularly engaged in conducting tests, furnishing inspection services or furnishing product evaluation or certification where such organization has been approved by the code official.

[A] BUILDING. Any structure intended for supporting or sheltering any occupancy.

[A] BUILDING OFFICIAL. The officer or other designated authority charged with the administration and enforcement of the *International Building Code*, or the building official's duly authorized representative.

CERTIFICATE OF COMPLETION. Written documentation that the project or work for which a permit was issued has been completed in conformance with requirements of this code.

[A] CODE OFFICIAL. The official designated by the jurisdiction to interpret and enforce this code, or the code official's authorized representative.

CRITICAL FIRE WEATHER. A set of weather conditions (usually a combination of low relative humidity and wind) whose effects on fire behavior make control difficult and threaten firefighter safety.

DEFENSIBLE SPACE. An area either natural or man-made, where material capable of allowing a fire to spread unchecked has been treated, cleared or modified to slow the rate and intensity of an advancing wildfire and to create an area for fire suppression operations to occur.

DRIVEWAY. A vehicular ingress and egress route that serves no more than two buildings or structures, not including accessory structures, or more than five dwelling units.

[BG] DWELLING. A building that contains one or two dwelling units used, intended or designed to be used, rented, leased, let or hired out to be occupied for living purposes.

[F] FIRE CHIEF. The chief officer or the chief officer's authorized representative of the fire department serving the jurisdiction.

FIRE FLOW CALCULATION AREA. The floor area, in square feet (square meters), used to determine the adequate water supply.

FIRE PROTECTION PLAN. A document prepared for a specific project or development proposed for the *wildland-urban interface area*. It describes ways to minimize and mitigate the fire problems created by the project or development, with the purpose of reducing impact on the community's fire protection delivery system.

FIRE WEATHER. Weather conditions favorable to the ignition and rapid spread of fire. In wildfires, this generally includes high temperatures combined with strong winds and low humidity. See "*Critical fire weather.*"

FIRE-RESISTANCE-RATED CONSTRUCTION. The use of materials and systems in the design and construction of a building or structure to safeguard against the spread of fire within a building or structure and the spread of fire to or from buildings or structures to the *wildland-urban interface area*.

[BG] FLAME SPREAD INDEX. A comparative measure, expressed as a dimensionless number, derived from visual measurements of the spread of flame versus time for a material tested in accordance with ASTM E84.

FUEL, HEAVY. Vegetation consisting of round wood 3 to 8 inches (76 to 203 mm) in diameter. See Fuel Models G, I, J, K and U described in Appendix D.

FUEL, LIGHT. Vegetation consisting of herbaceous plants and round wood less than $^1/_4$ inch (6.4 mm) in diameter. See Fuel Models A, C, E, L, N, P, R and S described in Appendix D.

FUEL, MEDIUM. Vegetation consisting of round wood $^1/_4$ to 3 inches (6.4 mm to 76 mm) in diameter. See Fuel Models B, D, F, H, O, Q and T described in Appendix D.

FUEL BREAK. An area, strategically located for fighting anticipated fires, where the native vegetation has been permanently modified or replaced so that fires burning into it can be more easily controlled. Fuel breaks divide fire-prone areas into smaller areas for easier fire control and to provide access for firefighting.

FUEL MODIFICATION. A method of modifying fuel load by reducing the amount of nonfire-resistive vegetation or altering the type of vegetation to reduce the fuel load.

FUEL MOSAIC. A *fuel modification* system that provides for the creation of islands and irregular boundaries to reduce the visual and ecological impact of *fuel modification*.

FUEL-LOADING. The oven-dry weight of fuels in a given area, usually expressed in pounds per acre (lb/a) (kg/ha). Fuel-loading may be referenced to fuel size or time-lag categories, and may include surface fuels or total fuels.

GREEN BELT. A *fuel break* designated for a use other than fire protection.

HAZARDOUS MATERIALS. As defined in the *International Fire Code*.

HEAVY TIMBER CONSTRUCTION. As described in the *International Building Code*.

IGNITION-RESISTANT BUILDING MATERIAL. A type of building material that resists ignition or sustained flaming combustion sufficiently so as to reduce losses from wildfire exposure of burning embers and small flames.

IGNITION-RESISTANT CONSTRUCTION, CLASS 1. A schedule of additional requirements for construction in *wildland-urban interface areas* based on extreme fire hazard.

IGNITION-RESISTANT CONSTRUCTION, CLASS 2. A schedule of additional requirements for construction in *wildland-urban interface areas* based on high fire hazard.

IGNITION-RESISTANT CONSTRUCTION, CLASS 3. A schedule of additional requirements for construction in *wildland-urban interface areas* based on moderate fire hazard.

LOG WALL CONSTRUCTION. A type of construction in which exterior walls are constructed of solid wood members and where the smallest horizontal dimension of each solid wood member is not less than 6 inches (152 mm).

MULTILAYERED GLAZED PANELS. Window or door assemblies that consist of two or more independently glazed panels installed parallel to each other, having a sealed air gap in between, within a frame designed to fill completely the window or door opening in which the assembly is intended to be installed.

NONCOMBUSTIBLE. As applied to building construction material means a material that, in the form in which it is used, is either one of the following:

1. Material of which no part will ignite and burn when subjected to fire. Any material conforming to ASTM E136 shall be considered noncombustible within the meaning of this section.
2. Material having a structural base of noncombustible material as defined in Item 1 above, with a surfacing material not over $^1/_8$ inch (3.2 mm) thick, which has a flame spread index of 50 or less. Flame spread index as used herein refers to a flame spread index obtained according to tests conducted as specified in ASTM E84 or UL 723.

"Noncombustible" does not apply to surface finish materials. Material required to be noncombustible for reduced clearances to flues, heating appliances or other sources of high temperature shall refer to material conforming to Item 1. No material shall be classified as noncombustible that is subject to increase in combustibility or flame spread index, beyond the limits herein established, through the effects of age, moisture or other atmospheric condition.

NONCOMBUSTIBLE ROOF COVERING. A roof covering consisting of any of the following:

1. Cement shingles or sheets.
2. Exposed concrete slab roof.
3. Ferrous or copper shingles or sheets.
4. Slate shingles.
5. Clay or concrete roofing tile.
6. *Approved* roof covering of *noncombustible* material.

[A] PEER REVIEW. An independent and objective technical review conducted by an approved third party.

[A] REGISTERED DESIGN PROFESSIONAL. An architect or engineer, registered or licensed to practice professional architecture or engineering, as defined by the statutory requirements of the professional registration laws of the state in which the project is to be constructed.

ROOF ASSEMBLY. A system designed to provide weather protection and resistance to design loads. The system consists of a *roof covering* and *roof deck* or a single component serving as both the *roof covering* and the *roof deck*. A roof assembly can include an underlayment, thermal barrier, ignition barrier, insulation or a vapor retarder.

ROOF COVERING. The covering applied to the *roof deck* for weather resistance, fire classification or appearance.

ROOF COVERING SYSTEM. See "*Roof assembly*."

ROOF DECK. The flat or sloped surface not including its supporting members or vertical supports.

SLOPE. The variation of terrain from the horizontal; the number of feet (meters) rise or fall per 100 feet (30 480 mm) measured horizontally, expressed as a percentage.

[A] STRUCTURE. That which is built or constructed.

[Z] SUBDIVISION. The division of a tract, lot or parcel of land into two or more lots, plats, sites or other divisions of land.

TREE CROWN. The primary and secondary branches growing out from the main stem, together with twigs and foliage.

UNENCLOSED ACCESSORY STRUCTURE. An accessory structure without a complete exterior wall system enclosing the area under roof or floor above.

WILDFIRE. An uncontrolled fire spreading through vegetative fuels, exposing and possibly consuming structures.

WILDLAND. An area in which development is essentially nonexistent, except for roads, railroads, power lines and similar facilities.

WILDLAND-URBAN INTERFACE AREA. That geographical area where structures and other human development meets or intermingles with wildland or vegetative fuels.

CHAPTER

3 WILDLAND-URBAN INTERFACE AREAS

User notes:

About this chapter: *Chapter 3 provides for the fundamental aspect of applying the code—the legal declaration and establishment of wildland-urban interface areas within the adopting jurisdiction by the local legislative body. The provisions cover area analysis and declaration based on findings of fact (located in Appendix E), mapping of the area, legal recordation of the maps with the local keeper of records and the periodic review and reevaluation of the declared areas on a regular basis. If needed, revisions can be directed by the legislative body of the jurisdiction.*

SECTION 301—GENERAL

301.1 Scope. The provisions of this chapter provide methodology to establish and record *wildland-urban interface areas* based on the findings of fact.

301.2 Objective. The objective of this chapter is to provide simple baseline criteria for determining *wildland-urban interface areas*.

SECTION 302—WILDLAND-URBAN INTERFACE AREA DESIGNATIONS

302.1 Declaration. The legislative body shall declare the *wildland-urban interface areas* within the jurisdiction. The *wildland-urban interface areas* shall be based on the findings of fact. The *wildland-urban interface area* boundary shall correspond to natural or man-made features.

302.2 Mapping. The *wildland-urban interface areas* shall be recorded on maps available for inspection by the public.

302.3 Review of wildland-urban interface areas. The *code official* shall reevaluate and recommend modification to the *wildland-urban interface areas* in accordance with Section 302.1 on a 3-year basis or more frequently as deemed necessary by the legislative body.

CHAPTER

4 WILDLAND-URBAN INTERFACE AREA REQUIREMENTS

User notes:

About this chapter: *Chapter 4 provides requirements that apply to all occupancies in the wildland-urban interface and pertain to all of the following:*

1. *Fire service access to the property that is to be protected, including fire apparatus access roads and off-road driveways.*
2. *Premises identification.*
3. *Key boxes to provide ready access to properties secured by gated roadways or other impediments to rapid fire service access.*
4. *Fire protection water supplies, including adequate water sources, pumper apparatus drafting sites, fire hydrant systems and system reliability.*
5. *Fire department access to equipment such as fire suppression equipment and fire hydrants.*
6. *Fire protection plans.*

SECTION 401—GENERAL

401.1 Scope. *Wildland-urban interface areas* shall be provided with emergency vehicle access and water supply in accordance with this chapter.

401.2 Objective. The objective of this chapter is to establish the minimum requirements for emergency vehicle access and water supply for buildings and structures located in the *wildland-urban interface areas*.

401.3 General safety precautions. General safety precautions shall be in accordance with this chapter. See also Appendix A.

SECTION 402—APPLICABILITY

402.1 Subdivisions. Subdivisions shall comply with Sections 402.1.1 and 402.1.2.

402.1.1 Access. New subdivisions, as determined by this jurisdiction, shall be provided with fire apparatus access roads in accordance with the *International Fire Code* and access requirements in accordance with Section 403.

402.1.2 Water supply. New subdivisions as determined by this jurisdiction shall be provided with water supply in accordance with Section 404.

402.2 Individual structures. Individual structures shall comply with Sections 402.2.1 and 402.2.2.

402.2.1 Access. Individual structures hereafter constructed or relocated into or within *wildland-urban interface areas* shall be provided with fire apparatus access in accordance with the *International Fire Code* and driveways in accordance with Section 403.2. Marking of fire protection equipment shall be provided in accordance with Section 403.5 and address markers shall be provided in accordance with Section 403.6.

402.2.2 Water supply. Individual structures hereafter constructed or relocated into or within *wildland-urban interface areas* shall be provided with a conforming water supply in accordance with Section 404.

Exceptions:

1. Structures constructed to meet the requirements for the class of ignition-resistant construction specified in Table 503.1 for a nonconforming water supply.
2. Buildings containing only private garages, carports, sheds and agricultural buildings with a floor area of not more than 600 square feet (56 m^2).

402.3 Existing conditions. Existing buildings shall be provided with address markers in accordance with Section 403.6. Existing roads and fire protection equipment shall be provided with markings in accordance with Sections 403.4 and 403.5, respectively.

SECTION 403—ACCESS

403.1 Restricted access. Where emergency vehicle access is restricted because of secured access roads or *driveways* or where immediate access is necessary for lifesaving or firefighting purposes, the *code official* is authorized to require a key box to be installed in an *approved* location. The key box shall be of a type *approved* by the *code official* and shall contain keys to gain necessary access as required by the *code official*.

403.2 Driveways. *Driveways* shall be provided where any portion of an exterior wall of the first story of a building is located more than 150 feet (45 720 mm) from a fire apparatus access road.

403.2.1 Dimensions. *Driveways* shall provide a minimum unobstructed width of 12 feet (3658 mm) and a minimum unobstructed height of 13 feet 6 inches (4115 mm).

403.2.2 Length. *Driveways* in excess of 150 feet (45 720 mm) in length shall be provided with turnarounds. *Driveways* in excess of 200 feet (60 960 mm) in length and less than 20 feet (6096 mm) in width shall be provided with turnouts in addition to turnarounds.

403.2.3 Service limitations. A *driveway* shall not serve in excess of five *dwelling* units.

Exception: Where such driveways meet the requirements for fire apparatus access roads in accordance with Section 503 of the *International Fire Code*.

403.2.4 Turnarounds. *Driveway* turnarounds shall have inside turning radii of not less than 30 feet (9144 mm) and outside turning radii of not less than 45 feet (13 716 mm). *Driveways* that connect with a road or roads at more than one point shall be considered as having a turnaround if all changes of direction meet the radii requirements for *driveway* turnarounds.

403.2.5 Turnouts. *Driveway* turnouts shall be an all-weather road surface not less than 10 feet (3048 mm) wide and 30 feet (9144 mm) long. *Driveway* turnouts shall be located as required by the *code official*.

403.2.6 Bridges. Vehicle load limits shall be posted at both entrances to bridges on *driveways* and private roads. Design loads for bridges shall be established by the *code official*.

403.3 Fire apparatus access road. Where required, fire apparatus access roads shall be all-weather roads with a minimum width of 20 feet (6096 mm) and a clear height of 13 feet 6 inches (4115 mm); shall be designed to accommodate the loads and turning radii for fire apparatus; and shall have a gradient negotiable by the specific fire apparatus normally used at that location within the jurisdiction. Dead-end roads in excess of 150 feet (45 720 mm) in length shall be provided with turnarounds as *approved* by the *code official*. An all-weather road surface shall be any surface material acceptable to the *code official* that would normally allow the passage of emergency service vehicles typically used to respond to that location within the jurisdiction.

403.4 Marking of roads. *Approved* signs or other *approved* notices shall be provided and maintained for access roads and *driveways* to identify such roads and prohibit the obstruction thereof.

403.4.1 Sign construction. Road identification signs and supports shall be of noncombustible materials. Signs shall have minimum 4-inch-high (102 mm) reflective letters with $^1/_2$-inch (12.7 mm) stroke on a contrasting 6-inch-high (152 mm) sign. Road identification signage shall be mounted at a height of 7 feet (2134 mm) from the road surface to the bottom of the sign.

403.5 Marking of fire protection equipment. Fire protection equipment and fire hydrants shall be clearly identified in a manner *approved* by the *code official* to prevent obstruction.

403.6 Address markers. Buildings shall have a permanently posted address, which shall be placed at each *driveway* entrance and be visible from both directions of travel along the road. In all cases, the address shall be posted at the beginning of construction and shall be maintained thereafter, and the address shall be visible and legible from the road on which the address is located.

403.6.1 Signs along one-way roads. Address signs along one-way roads shall be visible from both the intended direction of travel and the opposite direction.

403.6.2 Multiple addresses. Where multiple addresses are required at a single *driveway*, they shall be mounted on a single post, and additional signs shall be posted at locations where *driveways* divide.

403.6.3 Single-business sites. Where a roadway provides access solely to a single commercial or industrial business, the address sign shall be placed at the nearest road intersection providing access to that site.

403.7 Grade. The gradient for fire apparatus access roads and *driveways* shall not exceed the maximum *approved* by the *code official*.

SECTION 404—WATER SUPPLY

404.1 General. Where provided in order to qualify as a conforming water supply for the purpose of Table 503.1 or as required for new subdivisions in accordance with Section 402.1.2, an *approved* water source shall have an adequate water supply for the use of the fire protection service to protect buildings and structures from exterior fire sources or to suppress structure fires within the *wildland-urban interface area* of the jurisdiction in accordance with this section.

Exception: Buildings containing only private garages, carports, sheds and agricultural buildings with a floor area of not more than 600 square feet (56 m^2).

404.2 Water sources. The point at which a water source is available for use shall be located not more than 1,000 feet (305 m) from the building and be *approved* by the *code official*. The distance shall be measured along an unobstructed line of travel.

Water sources shall comply with the following:

1. Man-made water sources shall have a minimum usable water volume as determined by the adequate water supply needs in accordance with Section 404.5. This water source shall be equipped with an *approved* hydrant. The water level of the water source shall be maintained by rainfall, water pumped from a well, water hauled by a tanker or by seasonal high water of a stream or river. The design, construction, location, water level maintenance, access and access maintenance of man-made water sources shall be *approved* by the *code official*.
2. Natural water sources shall have a minimum annual water level or flow sufficient to meet the adequate water supply needs in accordance with Section 404.5. This water level or flow shall not be rendered unusable because of freezing. This water source shall have an *approved* draft site with an *approved* hydrant. Adequate water flow and rights for access to the water source shall be ensured in a form acceptable to the *code official*.

404.3 Draft sites. *Approved* draft sites shall be provided at natural water sources intended for use as fire protection for compliance with this code. The design, construction, location, access and access maintenance of draft sites shall be *approved* by the *code official*.

404.3.1 Access. The draft site shall have emergency vehicle access from an access road in accordance with Section 403.

404.3.2 Pumper access points. The pumper access point shall be either an emergency vehicle access area alongside a conforming access road or an *approved driveway* not longer than 150 feet (45 720 mm). Pumper access points and access *driveways* shall be designed and constructed in accordance with all codes and ordinances enforced by this jurisdiction. Pumper access points shall not require the pumper apparatus to obstruct a road or *driveway*.

404.4 Hydrants. Hydrants shall be designed and constructed in accordance with nationally recognized standards. The location and access shall be *approved* by the *code official*.

404.5 Adequate water supply. Adequate water supply shall be determined for purposes of initial attack and flame front control as follows:

1. One- and two-family dwellings. The required water supply for one- and two-family dwellings having a *fire flow calculation area* that does not exceed 3,600 square feet (334 m^2) shall be 1,000 gallons per minute (63.1 L/s) for a minimum duration of 30 minutes. The required water supply for one- and two-family dwellings having a *fire flow calculation area* in excess of 3,600 square feet (334 m^2) shall be 1,500 gallons per minute (95 L/s) for a minimum duration of 30 minutes.

   Exception: A reduction in required flow rate of 50 percent, as *approved* by the *code official*, is allowed where the building is provided with an *approved* automatic sprinkler system.

2. Buildings other than one- and two-family dwellings. The water supply required for buildings other than one- and two-family dwellings shall be as *approved* by the *code official* but shall be not less than 1,500 gallons per minute (95 L/s) for a duration of 2 hours.

   Exception: A reduction in required flow rate of up to 75 percent, as *approved* by the *code official*, is allowed where the building is provided with an *approved* automatic sprinkler system. The resulting water supply shall not be less than 1,500 gallons per minute (94.6 L/s).

404.6 Fire department. The water supply required by this code shall only be *approved* where a fire department, rated Class 9 or better in accordance with ISO Commercial Rating Service, 1995, is available.

404.7 Obstructions. Access to water sources required by this code shall be unobstructed at all times. The *code official* shall not be deterred or hindered from gaining immediate access to water source equipment, fire protection equipment or hydrants.

404.8 Identification. Water sources, draft sites, hydrants and fire protection equipment and hydrants shall be clearly identified in a manner *approved* by the *code official* to identify location and to prevent obstruction by parking and other obstructions.

404.9 Testing and maintenance. Water sources, draft sites, hydrants and other fire protection equipment required by this code shall be subject to periodic tests as required by the *code official*. Such equipment installed under the provisions of this code shall be maintained in an operative condition at all times and shall be repaired or replaced where defective. Additions, repairs, alterations and servicing of such fire protection equipment and resources shall be in accordance with *approved* standards.

404.10 Reliability. Water supply reliability shall comply with Sections 404.10.1 through 404.10.3.

404.10.1 Objective. The objective of this section is to increase the reliability of water supplies by reducing the exposure of vegetative fuels to electrically powered systems.

404.10.2 Clearance of fuel. *Defensible space* shall be provided around water tank structures, water supply pumps and pump houses in accordance with Section 603.

404.10.3 Standby power. Standby power shall be provided to pumps, controllers and related electrical equipment so that stationary water supply facilities within the *wildland-urban interface area* that are dependent on electrical power can provide the required water supply. The standby power system shall be in accordance with Section 2702 of the *International Building Code* and Section 1203 of the *International Fire Code*. The standby power source shall be capable of providing power for not less than 2 hours.

Exceptions:

1. Where *approved* by the *code official*, a standby power supply is not required where the primary power service to the stationary water supply facility is underground.
2. A standby power supply is not required where the stationary water supply facility serves not more than one single-family dwelling.

SECTION 405—FIRE PROTECTION PLAN

405.1 General. Where required by the *code official*, a *fire protection plan* shall be prepared.

405.2 Content. The plan shall be based on a site-specific wildfire risk assessment that includes considerations of location, topography, aspect, flammable vegetation, climatic conditions and fire history. The plan shall address water supply, access, building ignition and fire-resistance factors, fire protection systems and equipment, *defensible space* and vegetation management.

405.3 Cost. The cost of *fire protection plan* preparation and review shall be the responsibility of the applicant.

405.4 Plan retention. The *fire protection plan* shall be retained by the *code official*.

CHAPTER

5 SPECIAL BUILDING CONSTRUCTION REGULATIONS

User notes:

About this chapter: *Chapter 5 provides regulations that establish minimum standards for the location, design and construction of buildings and structures based on fire hazard severity in the wildland-urban interface.*

The construction provisions of Chapter 5 are intended to supplement the requirements of the International Building Code and address mitigation of the unique hazards posed to buildings by wildfire and to reduce the hazards of building fires spreading to wildland fuels. This is accomplished by requiring ignition-resistant construction materials based on the hazard severity of the building site. Construction features regulated include underfloor areas; roof coverings; eaves and soffits; gutters and downspouts; exterior walls, doors and windows; ventilation openings and accessory structures.

SECTION 501—GENERAL

501.1 Scope. Buildings and structures shall be constructed in accordance with the *International Building Code* and this code.

Exceptions:

1. Accessory structures not exceeding 120 square feet (11 m^2) in floor area where located not less than 50 feet (15 240 mm) from buildings containing habitable spaces.
2. Agricultural buildings not less than 50 feet (15 240 mm) from buildings containing habitable spaces.

501.2 Objective. The objective of this chapter is to establish minimum standards to locate, design and construct buildings and structures or portions thereof for the protection of life and property, to resist damage from wildfires, and to mitigate building and structure fires from spreading to wildland fuels. The minimum standards set forth in this chapter vary with the critical *fire weather*, slope and fuel type to provide increased protection, above the requirements set forth in the *International Building Code*, from the various levels of hazards.

501.3 Fire-resistance-rated construction. Where this code requires 1-hour *fire-resistance-rated construction*, the fire-resistance rating of building elements, components or assemblies shall be determined in accordance with the test procedures set forth in ASTM E119 or UL 263.

Exceptions:

1. The fire-resistance rating of building elements, components or assemblies based on the prescriptive designs prescribed in Section 721 of the *International Building Code.*
2. The fire-resistance rating of building elements, components or assemblies based on the calculation procedures in accordance with Section 722 of the *International Building Code.*

SECTION 502—FIRE HAZARD SEVERITY

502.1 General. The fire hazard severity of building sites for buildings hereafter constructed, modified or relocated into *wildland-urban interface areas* shall be established in accordance with Table 502.1. See also Appendix C.

TABLE 502.1—FIRE HAZARD SEVERITY

FUEL MODEL[b]	CRITICAL FIRE WEATHER FREQUENCY								
	≤ 1 day[a]			2 to 7 days[a]			≥ 8 days[a]		
	Slope (%)			Slope (%)			Slope (%)		
	≤ 40	41-60	≥ 61	≤ 40	41-60	≥ 61	≤ 40	41-60	≥ 61
Light fuel	M	M	M	M	M	M	M	M	H
Medium fuel	M	M	H	H	H	H	E	E	E
Heavy fuel	H	H	H	H	E	E	E	E	E

E = Extreme hazard.
H = High hazard.
M = Moderate hazard.
a. Days per annum.
b. Where required by the code official, fuel classification shall be based on the historical fuel type for the area.

502.2 Fire hazard severity reduction. The fire hazard severity identified in Table 502.1 is allowed to be reduced by implementing a vegetation management plan in accordance with Appendix B.

SECTION 503—IGNITION-RESISTANT CONSTRUCTION AND MATERIAL

Scan for Changes

73282ec

503.1 General. Buildings and structures hereafter constructed, modified or relocated into or within *wildland-urban interface areas* shall meet the construction requirements in accordance with Table 503.1. Class 1, Class 2 or Class 3, ignition-resistant construction shall be in accordance with Sections 504, 505 and 506, respectively. Materials required to be ignition-resistant materials shall comply with the requirements of Section 503.2.

TABLE 503.1—IGNITION-RESISTANT CONSTRUCTION[a]

DEFENSIBLE SPACE[c]	FIRE HAZARD SEVERITY					
	Moderate hazard		High hazard		Extreme hazard	
	Water supply[b]		Water supply[b]		Water supply[b]	
	Conforming[d]	Nonconforming[e]	Conforming[d]	Nonconforming[e]	Conforming[d]	Nonconforming[e]
Nonconforming	IR 2	IR 1	IR 1	IR 1 N.C.	IR 1 N.C.	Not Permitted
Conforming	IR 3	IR 2	IR 2	IR 1	IR 1	IR 1 N.C.
1.5 × Conforming	Not Required	IR 3	IR 3	IR 2	IR 2	IR 1

a. Access shall be in accordance with Section 403.
b. Subdivisions shall have a conforming water supply in accordance with Section 402.1.
IR 1 = Ignition-resistant construction in accordance with Section 504.
IR 2 = Ignition-resistant construction in accordance with Section 505.
IR 3 = Ignition-resistant construction in accordance with Section 506.
N.C. = Exterior walls shall have a fire-resistance rating of not less than 1 hour and the exterior surfaces of such walls shall be noncombustible. Usage of log wall construction is allowed.
c. Conformance based on Section 603.
d. Conformance based on Section 404.
e. A nonconforming water supply is any water system or source that does not comply with Section 404, including situations where there is no water supply for structure protection or fire suppression.

503.2 Ignition-resistant building material. Ignition-resistant building materials shall comply with any one of the requirements in Sections 503.2.1 through 503.2.4.

503.2.1 Noncombustible material. Material shall comply with the definition of *noncombustible* materials in Section 202.

503.2.2 Fire-retardant-treated wood. Fire-retardant-treated wood shall be identified for exterior use and shall meet the requirements of Section 2303.2 of the *International Building Code*.

503.2.3 Fire-retardant-treated wood roof coverings. Roof assemblies containing fire-retardant-treated wood shingles and shakes shall comply with the requirements of Section 1505.6 of the *International Building Code* and shall be classified as Class A roof assemblies as required in Section 1505.2 of the *International Building Code*.

503.2.4 Ignition-resistant building material. Material shall be tested on the front and back faces in accordance with the extended ASTM E84 or UL 723 test, for a total test period of 30 minutes, or with the ASTM E2768 test. The materials shall bear identification showing the fire test results. Panel products shall be tested with a ripped or cut longitudinal gap of $^{1}/_{8}$ inch (3.2 mm). The materials, when tested in accordance with the test procedures set forth in ASTM E84 or UL 723 for a test period of 30 minutes, or with ASTM E2768, shall comply with Sections 503.2.4.1 through 503.2.4.3.

Exception: Materials composed of a combustible core and a noncombustible exterior covering made from either aluminum at a minimum 0.019 inch (0.48 mm) thickness or corrosion-resistant steel at a minimum 0.0149 inch (0.38 mm) thickness shall not be required to be tested with a ripped or cut longitudinal gap.

503.2.4.1 Flame spread. The material shall exhibit a flame spread index not exceeding 25.

503.2.4.2 Flame front. The material shall exhibit a flame front that does not progress more than 10 feet 6 inches (3200 mm) beyond the centerline of the burner at any time during the test.

503.2.4.3 Weathering. Ignition-resistant building materials shall maintain their performance in accordance with this section under conditions of use. The materials shall meet the performance requirements for weathering (including exposure to temperature, moisture and ultraviolet radiation) contained in Sections 503.2.4.3.1 through 503.2.4.3.3, as applicable to the materials and conditions of use.

503.2.4.3.1 Evaluation requirements for weathering. Fire-retardant-treated wood, wood-plastic composite materials and plastic lumber materials shall be evaluated after weathering in accordance with Method A "Test Method for Accelerated Weathering of Fire-Retardant-Treated Wood for Fire Testing" in ASTM D2898.

503.2.4.3.2 Wood-plastic composite materials. Wood-plastic composite materials shall also demonstrate acceptable fire performance after weathering by the following procedure: first testing in accordance with ASTM E1354 at an incident heat flux of 50 kW/m^2 in the horizontal orientation, then weathering in accordance with ASTM D7032 and then retesting in accordance with ASTM E1354 and exhibiting an increase of no more than 10 percent in peak rate of heat release when compared to the peak heat release rate of the nonweathered material.

503.2.4.3.3 Plastic lumber materials. Plastic lumber materials shall also demonstrate acceptable fire performance after weathering by the following procedure: first testing in accordance with ASTM E1354 at an incident heat flux of 50 kW/m^2 in the horizontal orientation, then weathering in accordance with ASTM D6662 and then retesting in accordance with ASTM E1354 and exhibiting an increase of no more than 10 percent in peak rate of heat release when compared to the peak heat release rate of the nonweathered material.

SECTION 504—CLASS 1 IGNITION-RESISTANT CONSTRUCTION

504.1 General. Class 1 ignition-resistant construction shall be in accordance with Sections 504.2 through 504.11.

504.2 Roof assembly. Roofs shall have a *roof assembly* that complies with a Class A rating when tested in accordance with ASTM E108 or UL 790. For *roof assemblies* where the profile allows a space between the *roof covering* and *roof deck*, the space at the eave ends shall be firestopped to preclude entry of flames or embers or have one layer of 72-pound (32.4 kg) mineral-surfaced, nonperforated cap sheet complying with ASTM D3909 installed over the combustible *roof deck*.

Exceptions:

1. Class A *roof assemblies* include those with coverings of brick, masonry or an exposed concrete *roof deck*.
2. Class A *roof assemblies* also include ferrous or copper shingles or sheets, metal sheets and shingles, clay or concrete roof tile or slate installed on noncombustible decks or ferrous, copper or metal sheets installed without a *roof deck* on noncombustible framing.
3. Class A *roof assemblies* include minimum 16 oz/sq ft (0.0416 kg/m^2) copper sheets installed over combustible *roof decks*.

504.2.1 Roof valleys. Where provided, valley flashings shall be not less than 0.019 inch (0.48 mm) (No. 26 galvanized sheet gage) corrosion-resistant metal installed over a minimum 36-inch-wide (914 mm) underlayment consisting of one layer of 72-pound (32.4 kg) mineral-surfaced, nonperforated cap sheet complying with ASTM D3909 running the full length of the valley.

504.3 Protection of eaves. Eaves and soffits shall be protected on the exposed underside by ignition-resistant materials or by materials *approved* for not less than 1-hour *fire-resistance-rated construction*, 2-inch (51 mm) nominal dimension lumber, or 1-inch (25 mm) nominal fire-retardant-treated lumber or $^{3}/_{4}$-inch (19.1 mm) nominal fire-retardant-treated plywood, identified for exterior use and meeting the requirements of Section 2303.2 of the *International Building Code*. Fascias are required and shall be protected on the backside by ignition-resistant materials or by materials *approved* for not less than 1-hour *fire-resistance-rated construction* or 2-inch (51 mm) nominal dimension lumber.

504.4 Gutters and downspouts. Gutters and downspouts shall be constructed of *noncombustible* material. Gutters shall be provided with an *approved* means to prevent the accumulation of leaves and debris in the gutter.

504.5 Exterior walls. Exterior walls of buildings or structures shall be constructed with one of the following methods:

1. Materials *approved* for not less than 1-hour *fire-resistance-rated construction* on the exterior side.
2. *Approved noncombustible* materials.
3. Heavy timber or *log wall construction*.
4. Fire-retardant-treated wood on the exterior side. The fire-retardant-treated wood shall be labeled for exterior use and meet the requirements of Section 2303.2 of the *International Building Code*.
5. Ignition-resistant materials complying with Section 503.2 on the exterior side.

Such material shall extend from the top of the foundation to the underside of the roof sheathing.

504.5.1 Flashing. A minimum of 6 inches (152 mm) of metal flashing or noncombustible material applied vertically on the exterior of the wall shall be installed at the ground, decking and roof intersections.

504.6 Underfloor enclosure. Buildings or structures shall have underfloor areas enclosed to the ground with exterior walls in accordance with Section 504.5.

Exception: Complete enclosure shall not be required where the underside of exposed floors and exposed structural columns, beams and supporting walls are protected as required for exterior 1-hour *fire-resistance-rated construction* or *heavy timber construction* or fire-retardant-treated wood. The fire-retardant-treated wood shall be labeled for exterior use and meet the requirements of Section 2303.2 of the *International Building Code*.

504.7 Appendages and projections. *Unenclosed accessory structures* attached to buildings with habitable spaces and projections, such as decks, shall be not less than 1-hour fire-resistance-rated construction, *heavy timber construction* or constructed of one of the following:

1. *Approved noncombustible* materials.
2. Fire-retardant-treated wood identified for exterior use and meeting the requirements of Section 2303.2 of the *International Building Code*.

3. Ignition-resistant building materials in accordance with Section 503.2.

Exception: Coated materials shall not be used as the walking surface of decks.

504.7.1 Underfloor areas. Where the attached structure is located and constructed so that the structure or any portion thereof projects over a descending slope surface greater than 10 percent, the area below the structure shall have underfloor areas enclosed to within 6 inches (152 mm) of the ground, with exterior wall construction in accordance with Section 504.5.

504.8 Exterior glazing. Exterior windows, window walls and glazed doors, windows within exterior doors, and skylights shall be tempered glass, *multilayered glazed panels*, glass block or have a fire protection rating of not less than 20 minutes.

504.9 Exterior doors. Exterior doors shall be *approved noncombustible* construction, solid-core wood not less than $1^{3}/_{4}$ inches thick (44 mm), or have a fire protection rating of not less than 20 minutes. Windows within doors and glazed doors shall be in accordance with Section 504.8.

Exception: Vehicle access doors.

504.10 Vents. Where provided, ventilation openings for enclosed attics, gable ends, ridge ends, under eaves and cornices, enclosed eave soffit spaces, enclosed rafter spaces formed where ceilings are applied directly to the underside of roof rafters, underfloor ventilation, foundations and crawl spaces, or any other opening intended to permit ventilation, either in a horizontal or vertical surface, shall be in accordance with Section 504.10.1 or Section 504.10.2 to resist building ignition from the intrusion of burning embers and flame through the ventilation openings.

504.10.1 Performance requirements. Ventilation openings shall be fully covered with listed vents, tested in accordance with ASTM E2886, to demonstrate compliance with all the following requirements:

1. There shall be no flaming ignition of the cotton material during the Ember Intrusion Test.
2. There shall be no flaming ignition during the Integrity Test portion of the Flame Intrusion Test.
3. The maximum temperature of the unexposed side of the vent shall not exceed 662°F (350°C).

504.10.2 Prescriptive requirements. Where provided, attic ventilation openings, foundation or underfloor vents, or other ventilation openings in vertical or horizontal surfaces and vents through roofs shall not exceed 144 square inches (0.0929 m^2) each. Such vents shall be covered with noncombustible corrosion-resistant mesh with openings not to exceed $^{1}/_{8}$ inch (3.2 mm) or shall be designed and approved to prevent flame or ember penetration into the structure.

504.10.3 Vent locations. Attic ventilation openings shall not be located in soffits, in eave overhangs, between rafters at eaves or in other overhang areas. Gable-end and dormer vents shall be located not less than 10 feet (3048 mm) from lot lines. Underfloor ventilation openings shall be located as close to grade as practical.

504.11 Detached accessory structures. Detached accessory structures located less than 50 feet (15 240 mm) from a building containing habitable space shall have exterior walls constructed with materials *approved* for not less than 1-hour *fire-resistance-rated construction*, heavy timber, *log wall construction*, or constructed with *approved noncombustible* materials or fire-retardant-treated wood on the exterior side. The fire-retardant-treated wood shall be labeled for exterior use and meet the requirements of Section 2303.2 of the *International Building Code*.

504.11.1 Underfloor areas. Where the detached structure is located and constructed so that the structure or any portion thereof projects over a descending slope surface greater than 10 percent, the area below the structure shall have underfloor areas enclosed to within 6 inches (152 mm) of the ground, with exterior wall construction in accordance with Section 504.5 or underfloor protection in accordance with Section 504.6.

Exception: The enclosure shall not be required where the underside of exposed floors and exposed structural columns, beams and supporting walls are protected as required for exterior 1-hour *fire-resistance-rated construction* or *heavy timber construction* or fire-retardant-treated wood on the exterior side. The fire-retardant-treated wood shall be labeled for exterior use and meet the requirements of Section 2303.2 of the *International Building Code*.

SECTION 505—CLASS 2 IGNITION-RESISTANT CONSTRUCTION

Scan for Changes

452b193

505.1 General. Class 2 ignition-resistant construction shall be in accordance with Sections 505.2 through 505.11.

505.2 Roof assembly. Roofs shall have a *roof assembly* that complies with not less than a Class A rating when tested in accordance with ASTM E108 or UL 790, or an *approved noncombustible roof covering*. For *roof assemblies* where the profile allows a space between the *roof covering* and *roof deck*, the space at the eave ends shall be firestopped to preclude entry of flames or embers, or have one layer of cap sheet complying with ASTM D3909 installed over the combustible *roof deck*.

505.2.1 Roof valleys. Where provided, valley flashings shall be not less than 0.019-inch (0.48 mm) (No. 26 galvanized sheet gage) corrosion-resistant metal installed over a minimum 36-inch-wide (914 mm) underlayment consisting of one layer of 72-pound (32.4 kg) mineral-surfaced, nonperforated cap sheet complying with ASTM D3909 running the full length of the valley.

505.3 Protection of eaves. Combustible eaves, fascias and soffits shall be enclosed with solid materials with a minimum thickness of $^{3}/_{4}$ inch (19 mm). Exposed rafter tails shall not be permitted unless constructed of heavy timber materials.

505.4 Gutters and downspouts. Gutters and downspouts shall be constructed of *noncombustible* material. Gutters shall be provided with an *approved* means to prevent the accumulation of leaves and debris in the gutter.

505.5 Exterior walls. Exterior walls of buildings or structures shall be constructed with one of the following methods:

1. Materials *approved* for not less than 1-hour *fire-resistance-rated construction* on the exterior side.
2. *Approved noncombustible* materials.
3. *Heavy timber* or *log wall construction*.
4. Fire-retardant-treated wood on the exterior side. The fire-retardant-treated wood shall be labeled for exterior use and meet the requirements of Section 2303.2 of the *International Building Code*.
5. Ignition-resistant materials on the exterior side.

Such material shall extend from the top of the foundation to the underside of the roof sheathing.

505.6 Underfloor enclosure. Buildings or structures shall have underfloor areas enclosed to the ground, with exterior walls in accordance with Section 505.5.

Exception: Complete enclosure shall not be required where the underside of exposed floors and exposed structural columns, beams and supporting walls are protected as required for exterior 1-hour *fire-resistance-rated construction* or *heavy timber construction* or fire-retardant-treated wood. The fire-retardant-treated wood shall be labeled for exterior use and meet the requirements of Section 2303.2 of the *International Building Code*.

505.7 Appendages and projections. *Unenclosed accessory structures* attached to buildings with habitable spaces and projections, such as decks, shall be not less than 1-hour *fire-resistance-rated construction*, *heavy timber construction* or constructed of one of the following:

1. *Approved noncombustible* materials.
2. Fire-retardant-treated wood identified for exterior use and meeting the requirements of Section 2303.2 of the *International Building Code*.
3. Ignition-resistant building materials in accordance with Section 503.2.

Exception: Coated materials shall not be used as the walking surface of decks.

505.7.1 Underfloor areas. Where the attached structure is located and constructed so that the structure or any portion thereof projects over a descending slope surface greater than 10 percent, the area below the structure shall have underfloor areas enclosed to within 6 inches (152 mm) of the ground, with exterior wall construction in accordance with Section 505.5.

505.8 Exterior glazing. Exterior windows, window walls and glazed doors, windows within exterior doors, and skylights shall be tempered glass, *multilayered glazed panels*, glass block or have a fire protection rating of not less than 20 minutes.

505.9 Exterior doors. Exterior doors shall be *approved noncombustible* construction, solid core wood not less than $1^3/_4$ inches thick (45 mm), or have a fire protection rating of not less than 20 minutes. Windows within doors and glazed doors shall be in accordance with Section 505.8.

Exception: Vehicle access doors.

505.10 Vents. Where provided, ventilation openings for enclosed attics, gable ends, ridge ends, under eaves and cornices, enclosed eave soffit spaces, enclosed rafter spaces formed where ceilings are applied directly to the underside of roof rafters, underfloor ventilation, foundations and crawl spaces, or any other opening intended to permit ventilation, either in a horizontal or vertical surface, shall be in accordance with Section 505.10.1 or Section 505.10.2 to resist building ignition from the intrusion of burning embers and flame through the ventilation openings.

505.10.1 Performance requirements. Ventilation openings shall be fully covered with listed vents, tested in accordance with ASTM E2886, to demonstrate compliance with all the following requirements:

1. There shall be no flaming ignition of the cotton material during the Ember Intrusion Test.
2. There shall be no flaming ignition during the Integrity Test portion of the Flame Intrusion Test.
3. The maximum temperature of the unexposed side of the vent shall not exceed 662°F (350°C).

505.10.2 Prescriptive requirements. Where provided, attic ventilation openings, foundation or underfloor vents, or other ventilation openings in vertical or horizontal surfaces and vents through roofs shall not exceed 144 square inches (0.0929 m^2) each. Such vents shall be covered with noncombustible corrosion-resistant mesh with openings not to exceed 1/8 inch (3.2 mm) or shall be designed and approved to prevent flame or ember penetration into the structure.

505.10.3 Vent locations. Attic ventilation openings shall not be located in soffits, in eave overhangs, between rafters at eaves or in other overhang areas. Gable-end and dormer vents shall be located not less than 10 feet (3048 mm) from lot lines. Underfloor ventilation openings shall be located as close to grade as practical.

505.11 Detached accessory structures. Detached accessory structures located less than 50 feet (15 240 mm) from a building containing habitable space shall have exterior walls constructed with materials *approved* for not less than 1-hour *fire-resistance-rated construction*, heavy timber, *log wall construction*, or constructed with *approved noncombustible* materials or fire-retardant-treated wood on the exterior side. The fire-retardant-treated wood shall be labeled for exterior use and meet the requirements of Section 2303.2 of the *International Building Code*.

505.11.1 Underfloor areas. Where the detached *accessory structure* is located and constructed so that the structure or any portion thereof projects over a descending slope surface greater than 10 percent, the area below the structure shall have underfloor areas enclosed to within 6 inches (152 mm) of the ground, with exterior wall construction in accordance with Section 505.5 or underfloor protection in accordance with Section 505.6.

Exception: The enclosure shall not be required where the underside of exposed floors and exposed structural columns, beams and supporting walls are protected as required for exterior 1-hour *fire-resistance-rated construction* or heavy-timber construction or fire-retardant-treated wood on the exterior side. The fire-retardant-treated wood shall be labeled for exterior use and meet the requirements of Section 2303.2 of the *International Building Code*.

SECTION 506—CLASS 3 IGNITION-RESISTANT CONSTRUCTION

Scan for Changes

00ef222

506.1 General. Class 3 ignition-resistant construction shall be in accordance with Sections 506.2 through 506.4.

506.2 Roof assembly. Roofs shall have a *roof assembly* that complies with not less than a Class B rating when tested in accordance with ASTM E108 or UL 790 or an *approved noncombustible roof covering*. For *roof assemblies* where the profile allows a space between the *roof covering* and *roof deck*, the space at the eave ends shall be firestopped to preclude entry of flames or embers, or have one layer of cap sheet complying with ASTM D3909 installed over the combustible *roof deck*.

506.2.1 Roof valleys. Where provided, valley flashings shall be not less than 0.019-inch (0.44 mm) (No. 26 galvanized sheet gage) corrosion-resistant metal installed over a minimum 36-inch-wide (914 mm) underlayment consisting of one layer of 72-pound (32.4 kg) mineral-surfaced, nonperforated cap sheet complying with ASTM D3909 running the full length of the valley.

506.3 Underfloor enclosure. Buildings or structures shall have underfloor areas enclosed to the ground with exterior walls.

Exception: Complete enclosure shall not be required where the underside of exposed floors and exposed structural columns, beams and supporting walls are protected as required for exterior 1-hour *fire-resistance-rated construction*, fire-retardant-treated wood or *heavy timber construction*. Fire-retardant-treated wood shall be labeled for exterior use and meet the requirements of Section 2303.2 of the *International Building Code*.

506.4 Gutters and downspouts. Gutters and downspouts shall be constructed of *noncombustible* material. Gutters shall be provided with an *approved* means to prevent the accumulation of leaves and debris in the gutter.

506.5 Vents. Where provided, attic ventilation openings, foundation or underfloor vents, or other ventilation openings in vertical exterior walls and vents through roofs shall not exceed 144 square inches (0.0929 m^2) each. Such vents shall be covered with noncombustible corrosion-resistant mesh with openings not to exceed $^1/_8$ inch (3.2 mm) or shall be designed and approved to prevent flame or ember penetration into the structure.

SECTION 507—REPLACEMENT OR REPAIR OF ROOF COVERINGS

507.1 General. The *roof covering* on buildings or structures in existence prior to the adoption of this code that are replaced or have 25 percent or more replaced in a 12-month period shall be replaced with a *roof covering* required for new construction based on the type of ignition-resistant construction specified in accordance with Section 503.

CHAPTER

6 FIRE PROTECTION REQUIREMENTS

User notes:

About this chapter: *Chapter 6 establishes minimum fire protection requirements to mitigate the hazards to life and property from fire in the wildland-urban interface. The chapter includes both design-oriented and prescriptive mitigation strategies to reduce the hazards of fire originating within a structure spreading to wildland and fire originating in wildland spreading to structures.*

Especially targeted for a systems-approach to fire protection are those new buildings that are deemed to be particularly hazardous under Chapter 5; these buildings are required to be sprinklered. Other hazard mitigation strategies include establishing around structures defensible space zones wherein combustible vegetation and trees are regulated and kept away from buildings and trees are located 10 feet crown-to-crown away from each other. Additional hazards that are dealt with in Chapter 6 include spark arrestors on chimneys and regulated storage of combustible materials, firewood and LP-gas.

SECTION 601—GENERAL

601.1 Scope. The provisions of this chapter establish general requirements for new and existing buildings, structures and premises located within *wildland-urban interface areas.*

601.2 Objective. The objective of this chapter is to establish minimum requirements to mitigate the risk to life and property from wildland fire exposures, exposures from adjacent structures and to prevent structure fires spreading to wildland fuels.

SECTION 602—AUTOMATIC SPRINKLER SYSTEMS

602.1 General. An *approved* automatic sprinkler system shall be installed in all occupancies in new buildings required to meet the requirements for Class 1 ignition-resistant construction in Chapter 5. The installation of the automatic sprinkler systems shall be in accordance with nationally recognized standards.

SECTION 603—DEFENSIBLE SPACE

603.1 Objective. Provisions of this section are intended to modify the fuel load in areas adjacent to structures to create a *defensible space.*

603.2 Fuel modification. Buildings or structures, constructed in compliance with the conforming *defensible space* category of Table 503.1, shall comply with the *fuel modification* distances contained in Table 603.2. For all other purposes the *fuel modification* distance shall be not less than 30 feet (9144 mm) or to the lot line, whichever is less. Distances specified in Table 603.2 shall be measured on a horizontal plane from the perimeter or projection of the building or structure as shown in Figure 603.2. Distances specified in Table 603.2 are allowed to be increased by the *code official* because of a site-specific analysis based on local conditions and the *fire protection plan.*

TABLE 603.2—REQUIRED DEFENSIBLE SPACE

WILDLAND-URBAN INTERFACE AREA	FUEL MODIFICATION DISTANCE (feet)[a]
Moderate hazard	30
High hazard	50
Extreme hazard	100

For SI: 1 foot = 304.8 mm.
a. Distances are allowed to be increased due to site-specific analysis based on local conditions and the fire protection plan.

FIGURE 603.2—MEASUREMENTS OF FUEL MODIFICATION DISTANCE

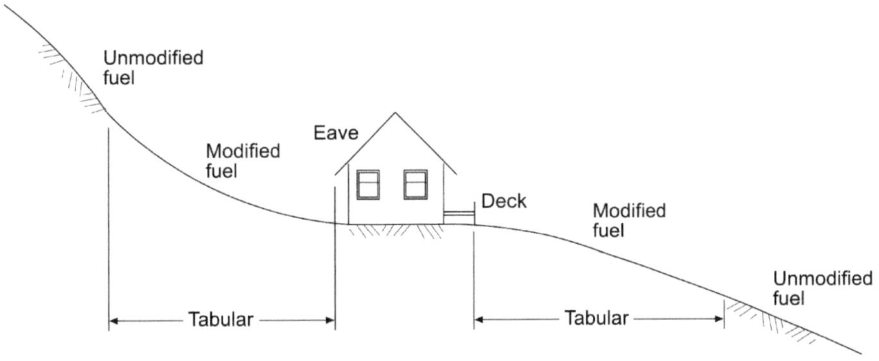

603.2.1 Responsible party. Persons owning, leasing, controlling, operating or maintaining buildings or structures requiring *defensible spaces* are responsible for modifying or removing nonfire-resistive vegetation on the property owned, leased or controlled by said person.

603.2.2 Trees. Trees are allowed within the *defensible space*, provided that the horizontal distance between crowns of adjacent trees and crowns of trees and structures, overhead electrical facilities or unmodified fuel is not less than 10 feet (3048 mm).

603.2.3 Ground cover. Deadwood and litter shall be regularly removed from trees. Where ornamental vegetative fuels or cultivated ground cover, such as green grass, ivy, succulents or similar plants are used as ground cover, they are allowed to be within the designated *defensible space*, provided that they do not form a means of transmitting fire from the native growth to any structure.

SECTION 604—MAINTENANCE OF DEFENSIBLE SPACE

604.1 General. *Defensible spaces* required by Section 603 shall be maintained in accordance with Section 604.

604.2 Modified area. Nonfire-resistive vegetation or growth shall be kept clear of buildings or structures, in accordance with Section 603, in such a manner as to provide a clear area for fire suppression operations.

604.3 Responsibility. Persons owning, leasing, controlling, operating or maintaining buildings or structures are responsible for maintenance of *defensible spaces*. Maintenance of the *defensible space* shall include modifying or removing nonfire-resistive vegetation and keeping leaves, needles and other dead vegetative material regularly removed from roofs of buildings and structures.

604.4 Trees. Tree crowns extending to within 10 feet (3048 mm) of any structure shall be pruned to maintain a minimum horizontal clearance of 10 feet (3048 mm). Tree crowns within the *defensible space* shall be pruned to remove limbs located less than 6 feet (1829 mm) above the ground surface adjacent to the trees.

604.4.1 Chimney clearance. Portions of tree crowns that extend to within 10 feet (3048 mm) of the outlet of a chimney shall be pruned to maintain a minimum horizontal clearance of 10 feet (3048 mm).

604.4.2 Deadwood removed. Deadwood and litter shall be regularly removed from trees.

SECTION 605—SPARK ARRESTORS

605.1 General. Chimneys serving fireplaces, barbecues, incinerators or decorative heating appliances in which solid or liquid fuel is used shall be provided with a spark arrestor. Spark arrestors shall be constructed of woven or welded wire screening of 12 USA standard gage wire (0.1046 inch) (2.66 mm) having openings not exceeding $^1/_2$ inch (12.7 mm).

605.2 Net free area. The net free area of the spark arrestor shall be not less than four times the net free area of the outlet of the chimney.

SECTION 606—LIQUEFIED PETROLEUM GAS INSTALLATIONS

606.1 General. The storage of liquefied petroleum gas (LP-gas) and the installation and maintenance of pertinent equipment shall be in accordance with the *International Fire Code* or, in the absence thereof, recognized standards.

606.2 Location of containers or tanks. LP-gas containers or tanks shall be located within the *defensible space* in accordance with the *International Fire Code*.

SECTION 607—STORAGE OF FIREWOOD AND COMBUSTIBLE MATERIALS

607.1 General. Firewood and combustible material shall not be stored in unenclosed spaces beneath buildings or structures, or on decks or under eaves, canopies or other projections or overhangs. Where required by the *code official*, storage of firewood and combustible material stored in the *defensible space* shall be located not less than 20 feet (6096 mm) from structures and separated from the crown of trees by a horizontal distance of not less than 15 feet (4572 mm).

607.2 Storage for off-site use. Firewood and combustible materials not for consumption on the premises shall be stored so as to not pose a hazard. See Appendix A.

CHAPTER
7 REFERENCED STANDARDS

User notes:

About this chapter: *This code contains numerous references to standards promulgated by other organizations that are used to provide requirements for materials and methods of construction. This chapter contains a comprehensive list of all standards that are referenced in this code. These standards, in essence, are part of this code to the extent of the reference to the standard.*

This chapter lists the standards that are referenced in various sections of this document. The standards are listed herein by the promulgating agency of the standard, the standard identification, the effective date and title, and the section or sections of this document that reference the standard.

ASTM *ASTM International, 100 Barr Harbor Drive, P.O. Box C700, West Conshohocken, PA 19428-2959*

D2898—10(2017): Standard Practice for Accelerated Weathering of Fire-Retardant-Treated Wood for Fire Testing
503.2

D3909/D3909M—14(2021): Standard Specification for Asphalt Roll Roofing (Glass Felt) Surfaced With Mineral Granules
504.2, 504.2.1, 505.2, 505.2.1, 506.2, 506.2.1

D6662—22: Standard Specification for Polyolefin-Based Plastic Lumber Decking Boards
503.2

D7032—2021: Standard Specification for Establishing Performance Ratings for Wood-Plastic Composite and Plastic Lumber Deck Boards, Stair Treads, Guards, and Handrails
503.2

E84—21a: Standard Test Method for Surface Burning Characteristics of Building Materials
202, 503.2

E108—20a: Standard Test Methods for Fire Tests of Roof Coverings
504.2, 505.2, 506.2

E119—20: Standard Test Methods for Fire Tests of Building Construction and Materials
501.3

E136—2022: Standard Test Method for Assessing Combustibility of Materials Using a Vertical Tube Furnace at 750 Degrees C
202

E1354—2022: Standard Test Method for Heat and Visible Smoke Release Rates for Materials and Products Using an Oxygen Consumption Calorimeter
503.2.4.3.2, 503.2.4.3.3

E2768—11(2018): Standard Test Method for Extended Duration Surface Burning Characteristics of Building Materials (30 min Tunnel Test)
503.2

E2886/E2886M—20: Standard Test Method for Evaluating the Ability of Exterior Vents to Resist the Entry of Embers and Direct Flame Impingement
504.10.1, 505.10.1

ICC *International Code Council, Inc., 200 Massachusetts Avenue, NW, Suite 250, Washington, DC 20001*

IBC—24: International Building Code®
105.1, 105.3, 106.3, 202, 404.10.3, 501.1, 501.2, 501.3, 503.2, 504.3, 504.5, 504.6, 504.7, 504.11, 505.5, 505.6, 505.7, 505.11

IFC—24: International Fire Code®
102.8, 105.1, 105.3, 202, 402.1.1, 402.2.1, 403.2.3, 404.10.3, 606.1, 606.2

IPMC—24: International Property Maintenance Code®
102.8

UL

UL LLC, 333 Pfingsten Road, Northbrook, IL 60062-2096

263—2011: Fire Tests of Building Construction and Materials—with Revisions through August 2021
501.3

723—2018: Standard for Test for Surface Burning Characteristics of Building Materials
202, 503.2

790—2004: Standard Test Methods for Fire Tests of Roof Coverings—with Revisions through October 2018
504.2, 505.2, 506.2

APPENDIX

A GENERAL REQUIREMENTS

The provisions contained in this appendix are not mandatory unless specifically referenced in the adopting ordinance.

User notes:

About this appendix: *Appendix A, while not part of the code, can become part of the code when specifically included in the adopting ordinance. Its purpose is to provide fire protection measures supplemental to those found in Chapter 6 to reduce the threat of wildfire in a wildland-urban interface area and improve the capability for controlling such fires. This appendix includes detailed requirements for vegetation control; the code official's authority to close wildland-interface areas in times of high fire danger; control of fires, fireworks usage and other sources of ignition; storage of hazardous materials and combustibles; bans on the dumping of waste materials and ashes and coals in wildland-urban interface areas; protection of pumps and water supplies; and limits on temporary uses within the wildland-urban interface area.*

Code development reminder: *Code change proposals to this appendix will be considered by the IFC Code Development Committee during the 2024 (Group A) Code Development Cycle.*

SECTION A101—GENERAL

A101.1 Scope. The provisions of this appendix establish general requirements applicable to new and existing properties located within *wildland-urban interface areas.*

A101.2 Objective. The objective of this appendix is to provide necessary fire protection measures to reduce the threat of wildfire in a *wildland-urban interface area* and improve the capability of controlling such fires.

SECTION A102—VEGETATION CONTROL

A102.1 General. Vegetation control shall comply with Sections A102.2 through A102.4.

A102.2 Clearance of brush or vegetative growth from roadways. The *code official* is authorized to require areas within 10 feet (3048 mm) on each side of portions of fire apparatus access roads and driveways to be cleared of nonfire-resistive vegetation growth.

Exception: Single specimens of trees, ornamental vegetative fuels or cultivated ground cover, such as green grass, ivy, succulents or similar plants used as ground cover, provided they do not form a means of readily transmitting fire.

A102.3 Clearance of brush and vegetative growth from electrical transmission and distribution lines. Clearance of brush and vegetative growth from electrical transmission and distribution lines shall be in accordance with Sections A102.3.1 through A102.3.2.3.

Exception: Sections A102.3.1 through A102.3.2.3 do not authorize persons not having legal right of entry to enter on or damage the property of others without consent of the owner.

A102.3.1 Support clearance. Persons owning, controlling, operating or maintaining electrical transmission or distribution lines shall have an *approved* program in place that identifies poles or towers with equipment and hardware types that have a history of becoming an ignition source, and provides a combustible free space consisting of a clearing of not less than 10 feet (3048 mm) in each direction from the outer circumference of such pole or tower during such periods of time as designated by the *code official*.

Exception: Lines used exclusively as telephone, telegraph, messenger call, alarm transmission or other lines classed as communication circuits by a public utility.

A102.3.2 Electrical distribution and transmission line clearances. Clearances between vegetation and electrical lines shall be in accordance with Sections A102.3.2.1 through A102.3.2.3.

A102.3.2.1 Trimming clearance. At the time of trimming, clearances not less than those established by Table A102.3.2.1 shall be provided. The radial clearances shown are minimum clearances that shall be established, at time of trimming, between the vegetation and the energized conductors and associated live parts.

Exception: The *code official* is authorized to establish minimum clearances different than those specified by Table A102.3.2.1 when evidence substantiating such other clearances is submitted to and *approved* by the *code official.*

TABLE A102.3.2.1—MINIMUM CLEARANCES BETWEEN VEGETATION AND ELECTRICAL LINES AT TIME OF TRIMMING

LINE VOLTAGE	MINIMUM RADIAL CLEARANCE FROM CONDUCTOR (feet)
2,400–72,000	4
72,001–110,000	6
110,001–300,000	10
300,001 or more	15

For SI: 1 foot = 304.8 mm.

A102.3.2.2 Minimum clearance to be maintained. Clearances not less than those established by Table A102.3.2.2 shall be maintained during such periods of time as designated by the *code official*. The site-specific clearance achieved, at time of pruning, shall vary based on species growth rates, the utility company-specific trim cycle, the potential line sway due to wind, line sag due to electrical loading and ambient temperature and the tree's location in proximity to the high voltage lines.

Exception: The *code official* is authorized to establish minimum clearances different than those specified by Table A102.3.2.2 when evidence substantiating such other clearances is submitted to and *approved* by the *code official*.

TABLE A102.3.2.2—MINIMUM CLEARANCES BETWEEN VEGETATION AND ELECTRICAL LINES TO BE MAINTAINED

LINE VOLTAGE	MINIMUM CLEARANCE (inches)
750–35,000	6
35,001–60,000	12
60,001–115,000	19
115,001–230,000	30.5
230,001–500,000	115

For SI: 1 inch = 25.4 mm.

A102.3.2.3 Electrical power line emergencies. During emergencies, the utility shall perform the required work to the extent necessary to clear the hazard. An emergency can include situations such as trees falling into power lines or trees in violation of Table A102.3.2.2.

A102.4 Correction of condition. The *code official* is authorized to give notice to the owner of the property on which conditions regulated by Section A102 exist to correct such conditions. If the owner fails to correct such conditions, the legislative body of the jurisdiction is authorized to cause the same to be done and make the expense of such correction a lien on the property where such condition exists.

SECTION A103—ACCESS RESTRICTIONS

A103.1 Restricted entry to public lands. The *code official* is authorized to determine and publicly announce when *wildland-urban interface areas* shall be closed to entry and when such areas shall again be opened to entry. Entry on and occupation of *wildland-urban interface areas*, except public roadways, inhabited areas or established trails and campsites that have not been closed during such time when the *wildland-urban interface area* is closed to entry, is prohibited.

Exceptions:

1. Residents and owners of private property within *wildland-urban interface areas* and their invitees and guests going to or being on their lands.
2. Entry, in the course of duty, by peace or police officers, and other duly authorized public officers, members of a fire department and members of the Wildland Firefighting Service.

A103.2 Trespassing on posted private property. Where the *code official* determines that a specific area within a *wildland-urban interface area* presents an exceptional and continuing fire danger because of the density of natural growth, difficulty of terrain, proximity to structures or accessibility to the public, such areas shall be restricted or closed until changed conditions warrant termination of such restriction or closure. Such areas shall be posted in accordance with Section A103.2.1.

A103.2.1 Signs. *Approved* signs prohibiting entry by unauthorized persons and referring to this code shall be placed on every closed area.

A103.2.2 Trespassing. Entering and remaining within areas closed and posted is prohibited.

Exception: Owners and occupiers of private or public property within closed and posted areas; their guests or invitees; authorized persons engaged in the operation and maintenance of necessary utilities such as electrical power, gas, telephone, water and sewer; and local, state and federal public officers and their authorized agents acting in the course of duty.

A103.3 Use of fire roads and defensible space. Motorcycles, motor scooters and motor vehicles shall not be driven or parked on, and trespassing is prohibited on, fire roads or *defensible space* beyond the point where travel is restricted by a cable, gate or sign, without the permission of the property owners. Vehicles shall not be parked in a manner that obstructs the entrance to a fire road or *defensible space*.

Exception: Public officers acting within their scope of duty.

A103.3.1 Obstructions. Radio and television aerials, guy wires thereto, and other obstructions shall not be installed or maintained on fire roads or *defensible spaces*, unless located 16 feet (4877 mm) or more above such fire road or *defensible space*.

A103.4 Use of motorcycles, motor scooters, ultralight aircraft and motor vehicles. Motorcycles, motor scooters, ultralight aircraft and motor vehicles shall not be operated within *wildland-urban interface areas*, without a permit by the *code official*, except on clearly established public or private roads. Permission from the property owner shall be presented when requesting a permit.

A103.5 Tampering with locks, barricades, signs and address markers. Locks, barricades, seals, cables, signs and address markers installed within *wildland-urban interface areas*, by or under the control of the *code official*, shall not be tampered with, mutilated, destroyed or removed.

A103.5.1 Gates, doors, barriers and locks. Gates, doors, barriers and locks installed by or under the control of the *code official* shall not be unlocked.

SECTION A104—IGNITION SOURCE CONTROL

A104.1 General. Ignition sources shall be controlled in accordance with Sections A104.2 through A104.10.

A104.2 Objective. Regulations in this section are intended to provide the minimum requirements to prevent the occurrence of wildfires.

A104.3 Clearance from ignition sources. Clearance between ignition sources and grass, brush or other combustible materials shall be maintained at not less than 30 feet (9144 mm).

A104.4 Smoking. Where required by the *code official*, signs shall be posted stating NO SMOKING. Persons shall not smoke within 15 feet (4572 mm) of combustible materials or nonfire-resistive vegetation.

Exception: Places of habitation or in the boundaries of established smoking areas or campsites as designated by the *code official.*

A104.5 Equipment and devices generating heat, sparks or open flames. Equipment and devices generating heat, sparks or open flames capable of igniting nearby combustibles shall not be used in *wildland-urban interface areas* without a permit from the *code official.*

Exception: Use of *approved* equipment within inhabited premises or designated campsites that are not less than 30 feet (9144 mm) from grass-, grain-, brush- or forest-covered areas.

A104.6 Fireworks. Fireworks shall not be used or possessed in *wildland-urban interface areas.*

Exception: Fireworks allowed by the *code official* under permit in accordance with the *International Fire Code* where not prohibited by applicable local or state laws, ordinances and regulations.

A104.6.1 Authority to seize. The *code official* is authorized to seize, take, remove or cause to be removed fireworks in violation of this section.

A104.7 Outdoor fires. Outdoor fires in *wildland-urban interface areas* shall comply with Sections A104.7.1 through A104.7.3.

A104.7.1 General. Persons shall not build, ignite or maintain any outdoor fire of any kind for any purpose in or on any *wildland-urban interface area*, except by the authority of a written permit from the *code official.*

Exception: Outdoor fires within inhabited premises or designated campsites where such fires are in a permanent barbecue, portable barbecue, outdoor fireplace, incinerator or grill and are not less than 30 feet (9144 mm) from any combustible material or nonfire-resistive vegetation.

A104.7.2 Permits. Permits shall incorporate such terms and conditions that will reasonably safeguard public safety and property. Outdoor fires shall not be built, ignited or maintained in or on hazardous fire areas under any of the following conditions:

1. When high winds are blowing.
2. When a person 17 years old or over is not present at all times to watch and tend such fire.
3. When a public announcement is made that open burning is prohibited.

A104.7.3 Restrictions. Persons shall not use a permanent barbecue, portable barbecue, outdoor fireplace or grill for the disposal of rubbish, trash or combustible waste material.

A104.8 Incinerators, outdoor fireplaces, permanent barbecues and grills. Incinerators, outdoor fireplaces, permanent barbecues and grills shall not be built, installed or maintained in *wildland-urban interface areas* without approval of the *code official.*

A104.8.1 Maintenance. Incinerators, outdoor fireplaces, permanent barbecues and grills shall be maintained in good repair and in a safe condition at all times. Openings in such appliances shall be provided with an *approved* spark arrestor, screen or door.

Exception: Where *approved* by the *code official*, unprotected openings in barbecues and grills necessary for proper functioning.

A104.9 Reckless behavior. The *code official* is authorized to stop any actions of a person or persons if the official determines that the action is reckless and could result in an ignition of fire or spread of fire.

A104.10 Planting vegetation under or adjacent to energized electrical lines. Vegetation that, at maturity, would grow to within 10 feet (3048 mm) of the energized conductors shall not be planted under or adjacent to energized power lines.

SECTION A105—CONTROL OF STORAGE

A105.1 General. In addition to the requirements of the *International Fire Code*, storage and use of the materials shall be in accordance with Sections A105.2 through A105.4.2.

A105.2 Hazardous materials. Hazardous materials in excess of 10 gallons (37.8 L) of liquid, 200 cubic feet (5.66 m^3) of gas, or 10 pounds (4.54 kg) of solids require a permit and shall comply with nationally recognized standards for storage and use.

A105.3 Explosives. Explosives shall not be possessed, kept, stored, sold, offered for sale, given away, used, discharged, transported or disposed of within *wildland-urban interface areas*, except by permit from the *code official*.

A105.4 Combustible materials. Outside storage of combustible materials such as, but not limited to, wood, rubber tires, building materials or paper products shall comply with the other applicable sections of this code and this section.

A105.4.1 Individual piles. Individual piles shall not exceed 5,000 square feet (465 m^2) of contiguous area. Piles shall not exceed 50,000 cubic feet (1416 m^3) in volume or 10 feet (3048 mm) in height.

A105.4.2 Separation. A clear space of not less than 40 feet (12 192 mm) shall be provided between piles. The clear space shall not contain combustible material or nonfire-resistive vegetation.

SECTION A106—DUMPING

A106.1 Waste material. Waste material shall not be placed, deposited or dumped in *wildland-urban interface areas* or in, on or along trails, roadways or highways or against structures in *wildland-urban interface areas*.

Exception: *Approved* public and *approved* private dumping areas.

A106.2 Ashes and coals. Ashes and coals shall not be placed, deposited or dumped in or on *wildland-urban interface areas*.

Exceptions:

1. In the hearth of an established fire pit, camp stove or fireplace.
2. In a noncombustible container with a tightfitting lid, which is kept or maintained in a safe location not less than 10 feet (3048 mm) from nonfire-resistive vegetation or structures.
3. Where such ashes or coals are buried and covered with 1 foot (305 mm) of mineral earth not less than 25 feet (7620 mm) from nonfire-resistive vegetation or structures.

SECTION A107—PROTECTION OF PUMPS AND WATER STORAGE FACILITIES

A107.1 General. The reliability of the water supply shall be in accordance with Sections A107.2 through A107.5.

A107.2 Objective. The intent of this section is to increase the reliability of water storage and pumping facilities and to protect such systems against loss from intrusion by fire.

A107.3 Fuel modification area. Water storage and pumping facilities shall be provided with a *defensible space* of not less than 30 feet (9144 mm) clear of nonfire-resistive vegetation or growth around and adjacent to such facilities.

Persons owning, controlling, operating or maintaining water storage and pumping systems requiring this *defensible space* are responsible for clearing and removing nonfire-resistive vegetation and maintaining the *defensible space* on the property owned, leased or controlled by said person.

A107.4 Trees. Portions of trees that extend to within 30 feet (9144 mm) of combustible portions of water storage and pumping facilities shall be removed.

A107.5 Protection of electrical power supplies. Where electrical pumps are used to provide the required water supply, such pumps shall be connected to a standby power source to automatically maintain electrical power in the event of power loss. The standby power source shall be capable of providing power for not less than 2 hours in accordance with Chapter 27 of the *International Building Code*, Section 1203 of the *International Fire Code* and NFPA 70.

Exception: A standby power source is not required where the primary power service to pumps is underground as *approved* by the *code official*.

SECTION A108—LAND USE LIMITATIONS

A108.1 General. Temporary fairs, carnivals, public exhibitions and similar uses must comply with all other provisions of this code in addition to enhanced ingress and egress requirements.

A108.2 Objective. The increased public use of land or structures in *wildland-urban interface areas* increases the potential threat to life safety. The provisions of this section are intended to reduce that threat.

A108.3 Permits. Temporary fairs, carnivals, public exhibitions or similar uses shall not be allowed in a designated *wildland-urban interface area*, except by permit from the *code official*.

Permits shall incorporate such terms and conditions that will reasonably safeguard public safety and property.

A108.4 Access roadways. In addition to the requirements in Section 403, access roadways shall be not less than 24 feet (7315 mm) wide and posted NO PARKING. Two access roadways shall be provided to serve the permitted use area.

Where required by the *code official* to facilitate emergency operations, *approved* emergency vehicle operating areas shall be provided.

SECTION A109—REFERENCED STANDARDS

A109.1 General. See Table A109.1 for standards that are referenced in various sections of this appendix. Standards are listed by the standard identification with the effective date, standard title and the section or sections of this appendix that reference the standard.

TABLE A109.1—REFERENCED STANDARDS

STANDARD ACRONYM	STANDARD NAME	SECTIONS HEREIN REFERENCED
IBC—24	*International Building Code*	A107.5
IFC—24	*International Fire Code*	A104.6, A105.1, A107.5
NFPA 70—23	*National Electrical Code*	A107.5

APPENDIX

B — VEGETATION MANAGEMENT PLAN

The provisions contained in this appendix are not mandatory unless specifically referenced in the adopting ordinance.

User notes:

About this appendix:	*Appendix B, while not part of the code, can become part of the code when specifically included in the adopting ordinance. Its purpose is to provide criteria for submitting vegetation management plans, specifying their content and establishing a criterion for considering vegetation management as being a fuel modification.*
Code development reminder:	*Code change proposals to this appendix will be considered by the IFC Code Development Committee during the 2024 (Group A) Code Development Cycle.*

SECTION B101—GENERAL

B101.1 Scope. Vegetation management plans shall be submitted to the *code official* for review and approval as part of the plans required for a permit.

B101.2 Plan content. Vegetation management plans shall describe all actions that will be taken to prevent a fire from being carried toward or away from the building. A vegetation management plan shall include the following information:

1. A copy of the site plan.
2. Methods and timetables for controlling, changing or modifying areas on the property. Elements of the plan shall include removal of slash, snags, vegetation that may grow into overhead electrical lines, other ground fuels, ladder fuels and dead trees, and the thinning of live trees.
3. A plan for maintaining the proposed fuel-reduction measures.

B101.3 Fuel modification. To be considered a *fuel modification* for purposes of this code, continuous maintenance of the clearance is required.

APPENDIX

C

FIRE HAZARD SEVERITY FORM

The provisions contained in this appendix are not mandatory unless specifically referenced in the adopting ordinance.

User notes:

About this appendix: *Appendix C, while not part of the code, can become part of the code (replacing Table 502.1) when specifically included in the adopting ordinance. Its purpose is to provide an alternative methodology to using Table 502.1 for analyzing the fire hazard severity of building sites using a pre-assigned value/scoring system for each feature that impacts the hazard level of a building site. Included in the evaluation are site access, types and management of vegetation, percentage of defensible space on the site, site topography, class of roofing and other construction materials used on the building (existing or to be constructed on the site), fire protection water supply, and whether utilities are installed above or below ground.*

Code development reminder: *Code change proposals to this appendix will be considered by the IFC Code Development Committee during the 2024 (Group A) Code Development Cycle.*

SECTION C101—FIRE HAZARD SEVERITY FORM

C101.1 Fire hazard severity form. Where adopted, Table C101.1 is permitted to be used as an alternative to Table 502.1 for analyzing the fire hazard severity of building sites.

TABLE C101.1—FIRE HAZARD SEVERITY FORM

A. Subdivision Design Points	
1. Ingress/Egress	
Two or more primary roads	1___
One road	3___
One-way road in, one-way road out	5___
2. Width of Primary Road	
20 feet (6096 mm) or more	1___
Less than 20 feet (6096 mm)	3___
3. Accessibility	
Road grade 5% or less	1___
Road grade more than 5%	3___
4. Secondary Road Terminus	
Loop roads, cul-de-sacs with an outside turning radius of 45 feet (13 716 mm) or greater	1___
Cul-de-sac turnaround	2___
Dead-end roads 200 feet (60 960 mm) or less in length	3___
Dead-end roads greater than 200 feet (60 960 mm) in length	5___
5. Street Signs	
Present	1___
Not present	3___
B. Vegetation (IWUIC Definitions)	
1. Fuel Types	
Light	1___
Medium	5___
Heavy	10___
2. Defensible Space	
70% or more of site	1___
30% or more, but less than 70% of site	10___
Less than 30% of site	20___
C. Topography	
8% or less	1___
More than 8%, but less than 20%	4___
20% or more, but less than 30%	7___

TABLE C101.1—FIRE HAZARD SEVERITY FORM—continued	
30% or more	10___
D. Roofing Material	
Class A Fire Rated	1___
Class B Fire Rated	5___
Class C Fire Rated	10___
Nonrated	20___
E. Fire Protection—Water Source	
500 GPM (1892.5 L/min) hydrant within 1,000 feet (304.8 m)	1___
Hydrant farther than 1,000 feet (304.8 m) or draft site	2___
Water source 20 min. or less, round trip	5___
Water source farther than 20 min., and 45 min. or less, round trip	7___
Water source farther than 45 min., round trip	10___
F. Existing Building Construction Materials	
Noncombustible siding/deck	1___
Noncombustible siding/combustible deck	5___
Combustible siding and deck	10___
G. Utilities (gas and/or electric)	
All underground utilities	1___
One underground, one above ground	3___
All above ground	5___
Total for Subdivision	
Moderate Hazard	40–59
High Hazard	60–74
Extreme Hazard	75+

APPENDIX D

FIRE DANGER RATING SYSTEM

This appendix is an excerpt from the National Fire Danger Rating System (NFDRS), 1978, United States Department of Agriculture Forest Service, General Technical Report INT-39, modified only to match I-Codes formatting with section or table numbers and titles, and is for information purposes and is not intended for adoption.

User notes:

About this appendix: *The fuel models included in Appendix D are only general descriptions because they represent all wildfire fuels from Florida to Alaska and from the East Coast to California.*

The National Fire Danger Rating System (NFDRS) is a set of computer programs and algorithms that allows land management agencies to estimate today's or tomorrow's fire danger for a given rating area. NFDRS characterizes fire danger by evaluating the approximate upper limit of fire behavior in a fire danger rating area during a 24-hour period based on fuels, topography and weather, or what is commonly called the fire triangle. Fire danger ratings are guides for initiating presuppression activities and selecting the appropriate level of initial response to a reported wildfire in lieu of detailed, site- and time-specific information.

Predicting the potential behavior and effects of wildland fire are essential tasks in fire management. Surface fire behavior and fire effects models and prediction systems are driven in part by fuelbed inputs such as load, bulk density, fuel particle size, heat content and moisture content. To facilitate use in models and systems, fuelbed inputs have been formulated into fuel models. A fuel model is a set of fuelbed inputs needed by a particular fire behavior or fire effects model. Different kinds of fuel models are used in fire spread models in a variety of fire behavior modeling systems. The fuel models in this appendix correlate with the light, medium and heavy fuel definitions found in Chapter 2 of the code.

SECTION D101—FUEL MODELS

D101.1 General. The Fuel Model Key is provided in Table D101.1. Fuel Models are described in Sections D101.1.1 through D101.1.20.

TABLE D101.1—FUEL MODEL KEY
I. Mosses, lichens and low shrubs predominate ground fuels. A. An overstory of conifers occupies more than one-third of the site: MODEL Q B. There is no overstory, or it occupies less than one-third of the site (tundra): MODEL S
II. Marsh grasses and/or reeds predominate: MODEL N

III. Grasses and/or forbs predominate.
A. There is an open overstory of conifer and/or hardwood trees: MODEL C
B. There is no overstory.
  1. Woody shrubs occupy more than one-third, but less than two-thirds of the site: MODEL T
  2. Woody shrubs occupy less than one-third of the site.
    a. The grasses and forbs are primarily annuals: MODEL A
    b. The grasses and forbs are primarily perennials: MODEL L

IV. Brush, shrubs, tree reproduction or dwarf tree species predominate.
A. Average height of woody plants is 6 feet or greater.
  1. Woody plants occupy two-thirds or more of the site.
    a. One-fourth or more of the woody foliage is dead.
      (1) Mixed California chaparral: MODEL B
      (2) Other types of brush: MODEL F
    b. Up to one-fourth of the woody foliage is dead: MODEL Q
    c. Little dead foliage: MODEL O
  2. Woody plants occupy less than two-thirds of the site: MODEL F
B. Average height of woody plants is less than 6 feet.
  1. Woody plants occupy two-thirds or more of the site.
    a. Western United States: MODEL F
    b. Eastern United States: MODEL O
  2. Woody plants occupy less than two-thirds but more than one-third of the site.
    a. Western United States: MODEL T
    b. Eastern United States: MODEL D
  3. Woody plants occupy less than one-third of the site.
    a. The grasses and forbs are primarily annuals: MODEL A
    b. The grasses and forbs are primarily perennials: MODEL L

TABLE D101.1—FUEL MODEL KEY—continued

V. Trees predominate.

- A. Deciduous broadleaf species predominate.
  - 1. The area has been thinned or partially cut, leaving slash as the major fuel component: MODEL K
  - 2. The area has not been thinned or partially cut.
    - a. The overstory is dormant; the leaves have fallen: MODEL E
    - b. The overstory is in full leaf: MODEL R
- B. Conifer species predominate.
  - 1. Lichens, mosses, and low shrubs dominate as understory fuels: MODEL Q
  - 2. Grasses and forbs are the primary ground fuels: MODEL C
  - 3. Woody shrubs and/or reproduction dominate as understory fuels.
    - a. The understory burns readily.
      - (1) Western United States: MODEL T
      - (2) Eastern United States:
        - (a) The understory is more than 6 feet tall: MODEL O
        - (b) The understory is less than 6 feet tall: MODEL D
    - b. The understory seldom burns: MODEL H
  - 4. Duff and litter, branchwood, and tree boles are the primary ground fuels.
    - a. The overstory is overmature and decadent; there is a heavy accumulation of dead tree debris: MODEL G
    - b. The overstory is not decadent; there is only a nominal accumulation of debris.
      - (1) The needles are 2 inches (51 mm) or more in length (most pines).
        - (a) Eastern United States: MODEL P
        - (b) Western United States: MODEL U
      - (2) The needles are less than 2 inches (51 mm) long: MODEL H

VI. Slash is the predominant fuel.

- A. The foliage is still attached; there has been little settling.
  - 1. The loading is 25 tons/acre (56.1 tons/ha) or greater: MODEL I
  - 2. The loading is less than 25 tons/acre (56.1 tons/ha) but more than 15 tons/acre (33.7 tons/ha): MODEL J
  - 3. The loading is less than 15 tons/acre (33.7 tons/ha): MODEL K
- B. Settling is evident; the foliage is falling off; grasses, forbs, and shrubs are invading the area.
  - 1. The loading is 25 tons/acre (56.1 tons/ha) or greater: MODEL J
  - 2. The loading is less than 25 tons/acre (56.1 tons/ha): MODEL K

D101.1.1 FUEL MODEL A. This fuel model represents western grasslands vegetated by annual grasses and forbs. Brush or trees may be present but are very sparse, occupying less than a third of the area. Examples of types where Fuel Model A should be used are cheatgrass and medusahead. Open pinyon-juniper, sagebrush-grass, and desert shrub associations may appropriately be assigned this fuel model if the woody plants meet the density criteria. The quantity and continuity of the ground fuels vary greatly with rainfall from year to year.

D101.1.2 FUEL MODEL B. Mature, dense fields of brush 6 feet (1829 mm) or more in height are represented by this fuel model. One-fourth or more of the aerial fuel in such stands is dead. Foliage burns readily. Model B fuels are potentially very dangerous, fostering intense, fast-spreading fires. This model is for California mixed chaparral generally 30 years or older. The F model is more appropriate for pure chamise stands. The B model may also be used for the New Jersey pine barrens.

D101.1.3 FUEL MODEL C. Open pine stands typify Model C fuels. Perennial grasses and forbs are the primary ground fuel but there is enough needle litter and branchwood present to contribute significantly to the fuel loading. Some brush and shrubs may be present but they are of little consequence. Situations covered by Fuel Model C are open, longleaf, slash, ponderosa, Jeffrey, and sugar pine stands. Some pinyon-juniper stands may qualify.

D101.1.4 FUEL MODEL D. This fuel model is specifically for the palmetto-gallberry understory-pine overstory association of the southeast coastal plains. It can also be used for the so-called "low pocosins" where Fuel Model O might be too severe. This model should only be used in the Southeast, because of a high moisture of extinction.

D101.1.5 FUEL MODEL E. Use this model after leaf fall for hardwood and mixed hardwood-conifer types where the hardwoods dominate. The fuel is primarily hardwood leaf litter. The oat-hickory types are best represented by Fuel Model E, but E is an acceptable choice for northern hardwoods and mixed forests of the Southeast. In high winds, the fire danger may be underrated because rolling and blowing leaves are not accounted for. In the summer after the trees have leafed out, Fuel Model E should be replaced by Fuel Model R.

D101.1.6 FUEL MODEL F. Fuel Model F is the only one of the 1972 NFDR System Fuel Models whose application has changed. Model F now represents mature closed chamise stands and oakbrush fields of Arizona, Utah and Colorado. It also applies to young, closed stands and mature, open stands of California mixed chaparral. Open stands of pinyon-juniper are represented; however, fire activity will be overrated at low wind speeds and where there is sparse ground fuels.

D101.1.7 FUEL MODEL G. Fuel Model G is used for dense conifer stands where there is a heavy accumulation of litter and downed woody material. Such stands are typically overmature and may also be suffering insect, disease, wind or ice damage-natural events that create a very heavy buildup of dead material on the forest floor. The duff and litter are deep, and much of the woody material is more than 3 inches (76 mm) in diameter. The undergrowth is variable, but shrubs are usually restricted to openings. Types meant to be represented by Fuel Model G are hemlock-Sitka spruce, Coast Douglas-fir, and wind-thrown or bug-killed stands of lodgepole pine and spruce.

D101.1.8 FUEL MODEL H. The short-needled conifers (white pines, spruces, larches and firs) are represented by Fuel Model H. In contrast to Model G fuels, Fuel Model H describes a healthy stand with sparse undergrowth and a thin layer of ground fuels. Fires in H fuels are typically slow spreading and are dangerous only in scattered areas where the downed woody material is concentrated.

D101.1.9 FUEL MODEL I. Fuel Model I was designed for clearcut conifer slash where the total loading of materials less than 6 inches (152 mm) in diameter exceeds 25 tons/acre (56.1 metric tons/ha). After settling and the fines (needles and twigs) fall from the branches, Fuel Model I will overrate the fire potential. For lighter loadings of clearcut conifer slash, use Fuel Model J, and for light thinnings and partial cuts where the slash is scattered under a residual overstory, use Fuel Model K.

D101.1.10 FUEL MODEL J. This model is complementary to Fuel Model I. It is for clearcuts and heavily thinned conifer stands where the total loading of materials less than 6 inches (152 mm) in diameter is less than 25 tons/acre (56.1 metric tons/ha). Again, as the slash ages, the fire potential will be overrated.

D101.1.11 FUEL MODEL K. Slash fuels from light thinnings and partial cuts in conifer stands are represented by Fuel Model K. Typically, the slash is scattered about under an open overstory. This model applies to hardwood slash and to southern pine clear-cuts where the loading of all fuels is less than 15 tons/acre (33.7 tons/ha).

D101.1.12 FUEL MODEL L This fuel model is meant to represent western grasslands vegetated by perennial grasses. The principal species are coarser and the loadings heavier than those in Model A fuels. Otherwise, the situations are very similar; shrubs and trees occupy less than one-third of the area. The quantity of fuel in these areas is more stable from year to year. In sagebrush areas, Fuel Model T may be more appropriate.

D101.1.13 FUEL MODEL N. This fuel model was constructed specifically for the saw-grass prairies of south Florida. It may be useful in other marsh situations where the fuel is coarse and reedlike. This model assumes that one-third of the aerial portion of the plants is dead. Fast-spreading, intense fires can occur even over standing water.

D101.1.14 FUEL MODEL O. The O fuel model applies to dense, brushlike fuels of the Southeast. O fuels, except for a deep litter layer, are almost entirely living, in contrast to B fuels. The foliage burns readily, except during the active growing season. The plants are typically over 6 feet (1829 mm) tall and are often found under an open stand of pine. The high pocosins of the Virginia, North and South Carolina coasts are the ideal of Fuel Model O. If the plants do not meet the 6-foot (1829 mm) criterion in those areas, Fuel Model D should be used.

D101.1.15 FUEL MODEL P. Closed, thrifty stands of long-needled southern pines are characteristic of P fuels. A 2- to 4-inch (51 to 102 mm) layer of lightly compacted needle litter is the primary fuel. Some small-diameter branchwood is present, but the density of the canopy precludes more than a scattering of shrubs and grass. Fuel Model P has the high moisture of extinction characteristic of the Southeast. The corresponding model for other long-needled pines is U.

D101.1.16 FUEL MODEL Q. Upland Alaskan black spruce is represented by Fuel Model Q. The stands are dense but have frequent openings filled with usually flammable shrub species. The forest floor is a deep layer of moss and lichens, but there is some needle litter and small-diameter branchwood. The branches are persistent on the trees, and ground fires easily reach into the tree crowns. This fuel model may be useful for jack pine stands in the Lake States. Ground fires are typically slow spreading, but a dangerous crowning potential exists.

D101.1.17 FUEL MODEL R. This fuel model represents the hardwood areas after the canopies leaf out in the spring. It is provided as the off-season substitute for E. It should be used during the summer in all hardwood and mixed conifer-hardwood stands where more than half of the overstory is deciduous.

D101.1.18 FUEL MODEL S. Alaskan or alpine tundra on relatively well-drained sites is the S fuel. Grass and low shrubs are often present, but the principal fuel is a deep layer of lichens and moss. Fires in these fuels are not fast spreading or intense, but are difficult to extinguish.

D101.1.19 FUEL MODEL T. The bothersome sagebrush-grass types of the Great Basin and the Intermountain West are characteristic of T fuels. The shrubs burn easily and are not dense enough to shade out grass and other herbaceous plants. The shrubs must occupy at least one-third of the site or the A or L fuel models should be used. Fuel Model T might be used for immature scrub oak and desert shrub associations in the West, and the scrub oak-wire grass type in the Southeast.

D101.1.20 FUEL MODEL U. Closed stands of western long-needled pines are covered by this model. The ground fuels are primarily litter and small branchwood. Grass and shrubs are precluded by the dense canopy but occur in the occasional natural opening. Fuel Model U should be used for ponderosa, Jeffrey, sugar pine, and red pine stands of the Lake States. Fuel Model P is the corresponding model for southern pine plantations.

APPENDIX

E FINDINGS OF FACT

This appendix is for informational purposes and is not intended for adoption.

User notes:

About this appendix: *Appendix E is an informational appendix that is intended to provide a methodology for presenting the findings of fact that are required by Chapter 3 when a jurisdiction defines and establishes a wildland-urban interface area that will be the subject of regulation by this code. The development of a written "findings of fact" that justifies designation of wildland-interface areas by local jurisdictions requires that a certain amount of research and analysis be conducted to support a written finding that is both credible and professional. In the context of adopting a supplemental document such as the wildland-urban interface declaration, the writing of these findings is essential in creating the maps and overlap needed to use their specific options.*

The purpose of this appendix is to provide an overview of how local officials could approach this process. Though other factors could be prioritized in findings based on the needs of a particular adopting jurisdiction, there are three essential phenomena commonly cited in adoption statutes: climate, topography and geography. The process demands a high level of professionalism to protect the jurisdiction's credibility in adopting more restrictive requirements. A superficial effort in preparing the findings of fact could jeopardize the proposed or adopted code restriction. Jurisdictions should devote a sufficient amount of time to draft the findings of fact to ensure that the facts are accurate, comprehensive and verifiable.

SECTION E101—GENERAL

E101.1 Objective. Originally, most fire and building codes were written and adopted at the local government level. As a result, there were many differences in code provisions from community to community. Local problems often resulted in unique code provisions that were appropriate to the local situation, but not of much use in other communities.

With the development of uniform and model codes and their subsequent adoption by state governments, the common features were applied everywhere. Once the basic provisions were codified into a format and structure that had appeal to both *code officials* and the builder-development community, their code became "minimum standards." The model codes were just that—documents setting minimum criteria that most communities could find acceptable but are not intended to solve every problem everywhere.

The model codes provide an option for exceptional situations requiring local modifications. Specific problems can be addressed by a specific process enabling the level of particular requirements to be adjusted. The solution that was commonly made available in the model adoption process was the development of a written "findings of fact" that justified modifications by local *code officials*. Many state codes identify a specific adoption process. This provision requires that a certain amount of research and analysis be conducted to support a written finding that is both credible and professional. In the context of adopting a supplemental document such as the wildland-urban interface provision, the writing of these findings is essential in creating the maps and overlap needed to use their specific options.

The purpose of this appendix is to provide an overview of how local *code officials* could approach this process.

SECTION E102—DEFINITIONS

E102.1 General. The following words and terms shall, for the purposes of this appendix, have the meanings shown herein. Refer to Chapter 2 for general definitions.

CLIMATE. The average course or condition of the weather at a particular place over a period of many years, as exhibited in absolute extremes, means and frequencies of given departures from these means (i.e., of temperature, wind velocity, precipitation and other weather elements).

GEOGRAPHY. "A science that deals with the earth and its life, especially the description of land, sea, air, and the distribution of plant and animal life including man and his industries with reference to the mutual relations of these diverse elements." Webster's Third New International Dictionary of the English Language, Unabridged.

INSURANCE SERVICES OFFICE (ISO). An agency that recommends fire insurance rates based on a grading schedule that incorporates evaluation of firefighting resources and capabilities.

TOPOGRAPHY. The configuration of landmass surface, including its relief (elevation) and the position of its natural and man-made features that affect the ability to cross or transit a terrain.

SECTION E103—CLIMATIC CONSIDERATIONS

E103.1 General. There are two types of climates: macro and micro. A macro climate affects an entire region and gives the area a general environmental context. A micro climate is a specific variation that could be related to the other two factors, topography and geography. A micro climate may cover a relatively small area or be able to encompass an entire community, as opposed to another community in the same county.

Climatic consideration should be given to the extremes, means and anomalies of the following weather elements:

1. Temperatures.

2. Relative humidities.
3. Precipitation and flooding conditions.
4. Wind speed and duration of periods of high velocity.
5. Wind direction.
6. Fog and other atmospheric conditions.

What is essential in creating a wildland-urban overlay are the data that suggest the existence of critical fire weather in the jurisdiction.

SECTION E104—TOPOGRAPHIC CONSIDERATIONS

E104.1 General. Topographic considerations should be given to the presence of the following topographical elements:

1. Elevation and ranges of elevation.
2. Location of ridges, drainages and escarpments.
3. Percent of grade (slope).
4. Location of roads, bridges and railroads.
5. Other topographical features, such as aspect exposure.

This information becomes an important part of creating analysis of *wildland-urban areas* because topography and slope are key elements (along with fuel type) that create the need for specific ignition-resistance requirements in this code.

SECTION E105—GEOGRAPHIC CONSIDERATIONS

E105.1 General. Geography should be evaluated to determine the relationship between man-made improvements (creating an exposure) and factors such as the following:

1. Fuel types, concentration in a mosaic and distribution of fuel types.
2. Earthquake fault zones.
3. Hazardous material routes.
4. Artificial boundaries created by jurisdictional boundaries.
5. Vulnerability of infrastructure to damage by climate and topographical concerns.

Fuel types are the final component of the findings that suggest the need for identifying *wildland-urban areas* in a jurisdiction. Review Appendix D for a brief description of the various fuel models that relate to the specific areas under evaluation.

SECTION E106—REPORTING THE FINDINGS

E106.1 General. After a person has researched a specific jurisdictional area, the facts should be incorporated into a written document that reflects how these facts relate to the *code official*'s specific needs. The following is an exhibit that incorporates one such report. It should be reviewed as an example of how a relationship can be drawn between specific facts, fire protection problems and specific code modifications. It should be noted that this is an example only.

EXHIBIT 1 — Findings

The [INSERT TITLE: ADMINISTRATOR] does herewith make findings that certain climatic, topographic or geological features exist in the [INSERT NAME: JURISDICTION], and that those features can, under certain circumstances, affect emergency services. Further, certain code amendments are made to the [INSERT: INTERNATIONAL FIRE CODE] and [INSERT: INTERNATIONAL BUILDING CODE] that are aimed at mitigating, to the extent possible, the impact of those features.

Finding 1

That the [INSERT NAME: JURISDICTION] is situated on the slopes of and at the base of the [INSERT: NAME OF MOUNTAINS]. Mountains, with drainages from the [INSERT: DIRECTION] portion of the district, including [INSERT: IDENTIFY LOCAL CREEKS/STREAMS/RIVERS], which, when flooded, could result in conditions rendering fire department vehicular traffic access unduly burdensome or impossible.

Further, the flood conditions described above carry the potential for overcoming the ability of the fire department to aid or assist in fire control, evacuations, rescues and the emergency task demands inherent in such situations. The potential for the aforementioned flooding conditions to result in limiting fire department emergency vehicular traffic, with resulting overtaxing fire department personnel, may further cause a substantial or total lack of protection against fire for the buildings and structures located within the jurisdiction.

The aforementioned conditions support the imposition of fire protection requirements greater than those set forth in the [INSERT: INTERNATIONAL BUILDING CODE OR INTERNATIONAL FIRE CODE].

Finding 2

That the [INSERT NAME: JURISDICTION] is situated near [INSERT: NUMBER OF FAULTS] major faults, each capable of generating earthquakes of significant magnitude. These are the [INSERT: NAME OF FAULTS]. These faults are subject to becoming active at any time; the [INSERT NAME: JURISDICTION] is particularly vulnerable to devastation should such an earthquake occur.

The potential effects of earthquake activity include isolating the [INSERT NAME: JURISDICTION] from the surrounding area and restricting or eliminating internal circulation due to the potential for collapsing of highway overpasses and underpasses, along with other bridges in the district, or an earthslide, and the potential for vertical movement rendering surface travel unduly burdensome or impossible.

Additional potential situations inherent in such an occurrence include loss of the [INSERT NAME: JURISDICTION] water sources; [INSERT: IDENTIFICATION OF LOCAL SOURCES] would be expected to suffer damage, along with the local reservoirs and water mains; broken natural gas mains causing structure and other fires; leakage of hazardous materials; the need for rescues from collapsed structures; and the rendering of first aid and other medical attention to large numbers of people.

The protection of human life and the preservation of property in the event of such an occurrence support the imposition of fire protection requirements greater than those set forth in the [INSERT: INTERNATIONAL BUILDING CODE OR INTERNATIONAL FIRE CODE].

Finding 3

That the [INSERT NAME: JURISDICTION] includes [INSERT: IDENTITY OF MAJOR TRANSPORTATION ROUTES]. [INSERT: IDENTITY OF ROUTE] is designated by the [INSERT NAME: JURISDICTION] as an approved transportation route for highly toxic and radioactive materials.

The potential for release or threatened release of a hazardous material along one of these routes is highly probable given the volume transported daily. Incidents of this nature will normally require all available emergency response personnel to prevent injury and loss of life and to prevent, as far as practicable, property loss. Emergency personnel responding to such aforementioned incidents may be unduly impeded and delayed in accomplishing an emergency response as a result of this situation, with the potential result of undue and unnecessary risk to the protection of life and public safety and, in particular, endangering residents and occupants in buildings or structures without the protection of automatic sprinklers.

The aforementioned problems support the imposition of fire protection requirements greater than those set forth in the [INSERT: INTERNATIONAL BUILDING CODE OR INTERNATIONAL FIRE CODE].

Finding 4

The seasonal climatic conditions during the late summer and fall create numerous serious difficulties regarding the control of and protection against fires in the [INSERT NAME: JURISDICTION].The hot, dry weather typical of this area in summer and fall, coupled with [INSERT: IDENTITY OF ADDITIONAL CLIMATIC CONDITIONS] frequently results in wildfires that threaten or could threaten the [INSERT NAME: JURISDICTION].

Although some code requirements, such as fire-resistive roof classification, have a direct bearing on building survival in a wildland fire situation, others, such as residential automatic sprinklers, may also have a positive effect. In dry climate on low humidity days, many materials are much more easily ignited. More fires are likely to occur and any fire, once started, can expand extremely rapidly. Residential automatic sprinklers can arrest a fire starting within a structure before the fire is able to spread to adjacent brush and structures.

Seasonal winds also have the potential for interfering with emergency vehicle access, delaying or making impossible fire responses, because of toppling of extensive plantings of [INSERT: TYPE OF TREES] trees. The trees are subject to uprooting in strong winds due to relatively small root bases compared to the tree itself.

The aforementioned problems support the imposition of fire protection requirements greater than those set forth in the [INSERT: INTERNATIONAL BUILDING CODE OR INTERNATIONAL FIRE CODE].

Finding 5

The [INSERT NAME: JURISDICTION] is a [INSERT: DESCRIBE TYPE OF REGION] and experiences water shortages from time to time. Those shortages can have a severely adverse effect on water availability for firefighting.

Fires starting in sprinklered buildings are typically controlled by one or two sprinkler heads, flowing as little as 13 gallons per minute (0.82 L/s) each.

Hose streams used by engine companies on well-established structure fires operate at about 250 gallons per minute (15.8 L/s) each, and the estimated water need for a typical residential fire is 1,250 to 1,500 gallons per minute (78.9 to 94.6 L/s), according to the Insurance Services Office.

Under circumstances such as earthquakes, when multiple fires start within the community, the limited water demands of residential automatic sprinklers would control and extinguish many fires before they spread from building to building. In such a disaster, water demands needed for conflagration firefighting probably would not be available.

The aforementioned problems support the imposition of fire protection requirements greater than those set forth in the [INSERT: INTERNATIONAL BUILDING CODE OR INTERNATIONAL FIRE CODE].

Finding 6

The topography of the [INSERT NAME: JURISDICTION] presents problems in delivery of emergency services, including fire protection. Hilly terrain has narrow, winding roads with little circulation, preventing rapid access and orderly evacuation. Much of these hills are covered with highly nonfire-resistive natural vegetation. In addition to access and evacuation problems, the terrain makes delivery of water extremely difficult. Some hill areas are served by water pump systems subject to failure in fire, high winds, earthquake and other power failure situations.

The aforementioned problems support the imposition of fire protection requirements greater than those set forth in the [INSERT: INTERNATIONAL BUILDING CODE OR INTERNATIONAL FIRE CODE].

SUMMARY

Efforts to produce comprehensive findings of fact cannot be underestimated. It is an essential step for fire protection professionals to take before risking the proposal to modify a model code with a requirement that is unique to that community. Done properly, a findings-of-fact document will not only support the adoption of a local modification, it may make it virtually impossible to ignore the need without creating a community consequence.

APPENDIX F

CHARACTERISTICS OF FIRE-RESISTIVE VEGETATION

This appendix is for information purposes and is not intended for adoption.

User notes:

About this appendix:	*Appendix F is an informational appendix provided for the convenience of the code user. It is simply a compilation of the eight characteristics of fire-resistive vegetation that can be used effectively within wildland-urban interface areas to reduce the likelihood of fire spread through vegetation.*

SECTION F101—GENERAL

F101.1 Characteristics of fire-resistive vegetation. All plants will burn under extreme fire weather conditions, such as drought. However, plants burn at different intensities and rates of consumption. Fire-resistive plants burn at a relatively low intensity, slow rates of spread and with short flame lengths. The following are characteristics of fire-resistive vegetation:

1. Growth with little or no accumulation of dead vegetation (either on the ground or standing upright).
2. Nonresinous plants (willow, poplar or tulip trees).
3. Low volume of total vegetation (for example, a grass area as opposed to a forest or shrub-covered land).
4. Plants with high live fuel moisture (plants that contain a large amount of water in comparison to their dry weight).
5. Drought-tolerant plants (deeply rooted plants with thick, heavy leaves).
6. Stands without ladder fuels (plants without small, fine branches and limbs between the ground and the canopy of overtopping shrubs and trees).
7. Plants requiring little maintenance (slow-growing plants that, when maintained, require little care).
8. Plants with woody stems and branches that require prolonged heating to ignite.

APPENDIX G

SELF-DEFENSE MECHANISM

This appendix is for information purposes and is not intended for adoption.

User notes:

About this appendix: *Appendix G is an information appendix that provides discussion of some elements of the proposed self-defense mechanisms and their role in enhancing the protection of exposed structures in the wildland-urban interface. To accept alternative self-defense mechanisms, the code official must carefully examine whether these devices will be in place at the time of an event and whether they will assist or actually complicate the defense of the structure by fire suppression forces if they are available.*

SECTION G101—GENERAL

G101.1 Identification of the problem. The *International Wildland-Urban Interface Code* establishes a set of minimum standards to reduce the loss of property from wildfire. The purpose of these standards is to prevent wildfire spreading from vegetation to a building. Frequently, proposals are made by property or landowners of buildings located in the wildland-urban interface to consider other options and alternatives instead of meeting these minimum standards. This appendix chapter provides discussion of some elements of the proposed self-defense mechanisms and their role in enhancing the protection of exposed structures.

G101.2 Structural survivability. Various stages of assault occur as a building is exposed to a wildland-urban fire. Ashes are cast in front of a fire out of a smoke or convection column, which can result in secondary ignitions. Heavier embers that have more body weight and may contain more heat to serve as sources of ignition follow. Finally, the actual intrusion of a flame front and the radiant heat flux can expose combustibles outside of a building and the exterior structure of a building to various levels of radiant heat. A study revealed that the actual exposure of a building to the flame front by the perimeter of the fire was usually less than 6 minutes. However, the exposure to the forms of other materials that can result in proliferation of other ignitions can vary, depending on wind, topography and fuel conditions.

To enhance structural survivability, the self-defense mechanisms must, first, do everything possible to prevent the ignition of materials from objects that are cast in front of the fire and, second, they must withstand the assault of the fire on the structure to prevent flames from penetrating into the building and resulting in an interior fire. There are considerable problems in achieving both of these objectives using some of the proposed alternative forms of protection such as the lack of definitive standards for self-defense mechanisms on the exterior of buildings. Although fire service has done considerable research into the evaluation of technology, such as smoke detectors, fire alarms, and interior sprinkler systems, very limited amount of study has been done on exterior sprinkler systems.

All forms of fire protection are classified as either active or passive. Active fire protection is taking specific action to control the fire in some manner. Passive fire protection uses resistance to ignition or provides some form of warning that allows other action to be taken. These two classifications of self-defense mechanisms create different problems with regard to being accepted as alternatives for building construction. Furthermore, certain self-defense mechanisms must be built in during new construction, and others may only be capable of being added as a retrofit to existing structures. As a matter of public policy, most *code officials* are reluctant to accept passive fire protection as an equivalent to a construction requirement, but are also reluctant to accept active fire protection systems that require intervention by suppression personnel.

The unequal distribution of self-defense mechanisms within a specific neighborhood poses another problem. If an individual is granted a waiver or exemption on the basis of putting in a nonmandated self-defense mechanism, and the neighbors to either side choose not to do so, or are not given the same options, there is a potential operational problem.

G101.3 Alternative concepts. Sections G101.3.1 through G101.3.6 provide consideration of several alternative self-defense mechanisms.

G101.3.1 Exterior sprinkler systems. Currently, there is no nationally accepted standard for the design and installation of exterior fire sprinkler systems. Interior sprinkler systems are regulated by nationally recognized standards that have specific requirements. However, exterior sprinkler systems lack such uniformity. What is generally proposed is a type of sprinkler system, placed on the roofs or eaves of a building, whose primary purpose is to wet down the roof. These types of systems can be activated either manually or automatically. However, the contemporary thought on exterior sprinkler systems is that if the roof classification is of sufficient fire resistance, exterior sprinklers are of little or no value. Another option and alternative with exterior sprinklers is to use them to improve the relative humidity and fuel moisture in the *defensible space*. In this case, the exterior sprinkler is not used to protect the structure as much as it attempts to alter the fuel situation. However, studies do not support the idea that merely spraying water into the air in the immediate vicinity of a rapidly advancing wildland-urban fire does much good. Clearly, irrigation systems that keep plants healthy and fire-resistive plants that resist convection and radiated heat can accomplish the same purpose.

G101.3.2 Alternative water supply systems for exposure protection. Pools and spas are often offered as an alternative water source for fire departments. These water sources must be reliable and able to be accessed to be of any use by fire protection forces. Accessibility means that the fire department must be able to withdraw the water without having to go through extraordinary measures such as knocking down fences or having to set up drafting situations. Designs have been created to put liquid- or gas-fueled pumps or gravity valves on pools and spas to allow fire departments to access these water systems. A key vulnerability to the use of these alternative water systems is loss of electrical power. When the reliability of a water system depends on external power sources, it cannot be relied upon by fire fighters to be available in a worst-case scenario.

G101.3.3 Class A foam systems. One technology is Class A foam devices. These are devices that allow a homeowner to literally coat the exterior of their house with a thick layer of foam that prevents the penetration of embers and radiant heat to the structure. Experiments in various wildland fire agencies advocate foaming houses in advance of fire and flame fronts. To be accepted by the *code official*, the Class A foam system should pass rigorous scrutiny with regard to the manner and needs in which it is activated, the ways and means in which it is properly maintained, and a ways and means to test the system for its operational readiness during hiatus between emergencies.

G101.3.4 Enhanced exterior fire protection. This alternative method would increase the degree of fire resistance on the exterior of a building. This is most often an alternative recommended as a retroactive application when individual properties cannot achieve adequate *defensible space* on the exterior of a building. Normally, fire resistance and building scenarios are concerned with containing a fire. Fire-resistance ratings within building design ensures resistance to a fire for the specified time to compartmentalize the building's interior.

To improve fire resistance on the exterior of the structure, the primary emphasis is on preventing intrusion into the building. This means protection of apertures and openings that may or may not be required to have any degree of fire resistance by accepted building codes. The option that is available here is for individuals to provide coverage in the form of shutters or closures to these areas, which, along with maintenance of combustible-free perimeters, can often prevent intrusion.

There are obvious limitations to this alternative. First and foremost is the means of adequately evaluating the proposed fire resistance of any given assembly. Testing techniques to determine fire resistance for such objects as drywall and other forms of construction may not be applicable to exterior application. Nonetheless, *code officials* should determine the utility of a specific fire resistance proposal by extrapolating conservatively.

G101.3.5 Shelter in place. Developments in the wildland-urban interface may be designed to allow occupants to "shelter in place." Use of this design alternative should include ignition-resistant construction, access, water supply, automatic sprinkler systems, provisions for and maintenance of *defensible space*, and a *Fire Protection Plan*.

A *Fire Protection Plan* describes ways to minimize the fire problems created by a specific project or development. The purpose for the *Fire Protection Plan* is to reduce the burden and impact of the project or development on the community's fire protection delivery system. The plan may utilize components of land use, building construction, vegetation management and other design techniques and technologies. It should include specific mitigation measures consistent with the unique problems resulting from the location, topography, geology, flammable vegetation and climate of the proposed site. The plan shall be consistent with this code, and approved by the fire *code official*. The cost of preparation and review is to be borne by the project or development proponent.

G101.3.6 Building location. The location of a new building within lot lines should be considered as it relates to topography and fire behavior. Buildings located in natural chimneys, such as narrow canyons and saddles, are especially fire prone because winds are funneled into these areas and eddies are created. Buildings located on narrow ridges without setbacks may be subjected to increased flame and convective heat exposure from a fire advancing from below. Stone or masonry walls can act as heat shields and deflect the flames. Swimming pools and rated or noncombustible decks and patios can be used to create a setback, decreasing the exposure to the structure. Attic and under-floor vents, picture windows and sliding glass doors should not face possible corridors due to the increased risk of flame or ember penetration.

G101.4 Conclusion. The purpose of the *International Wildland-Urban Interface Code* is to establish minimum standards that prevent the loss of structures, even if fire department intervention is absent. To accept alternative self-defense mechanisms, the *code official* must carefully examine whether these devices will be in place at the time of an event and whether they will assist or actually complicate the defense of the structure by fire suppression forces if they are available.

The best alternative to having a building comply with all of the provisions of this code is to remove sources of fuel. This is closely paralleled by excellent housekeeping between the vegetation and the structure. Alternative ways of achieving each of these goals can and should be considered after scrutiny by appropriately credentialed and qualified fire protection personnel.

APPENDIX

H

INTERNATIONAL WILDLAND-URBAN INTERFACE CODE FLOWCHART

This appendix is for information purposes and is not intended for adoption.

User notes:

About this appendix: *Appendix H is an informational appendix that is based on the "Decision Tree" concept and is intended to provide the code official with a graphical representation of how this code is to be applied in an orderly manner.*

SECTION H101—GENERAL

H101.1 Flowchart. The flowchart shown in Figures H101.1(1) through H101.1(4) is based on the "Decision Tree" concept and is intended to provide the *code official* with a graphical representation of how this code is to be applied.

FIGURE H101.1(1)—DETERMINATION OF REQUIREMENTS

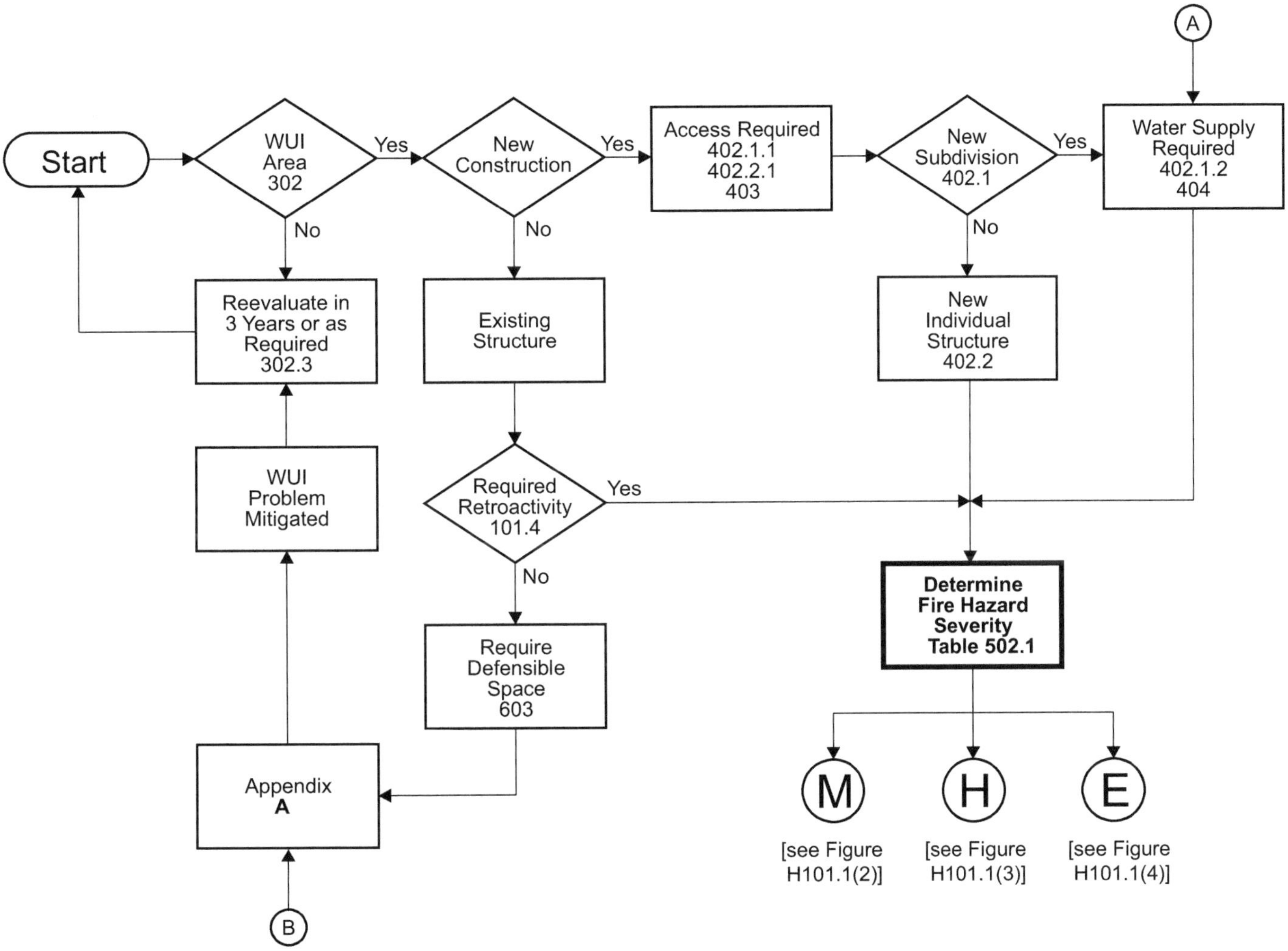

FIGURE H101.1(2)—MODERATE HAZARD

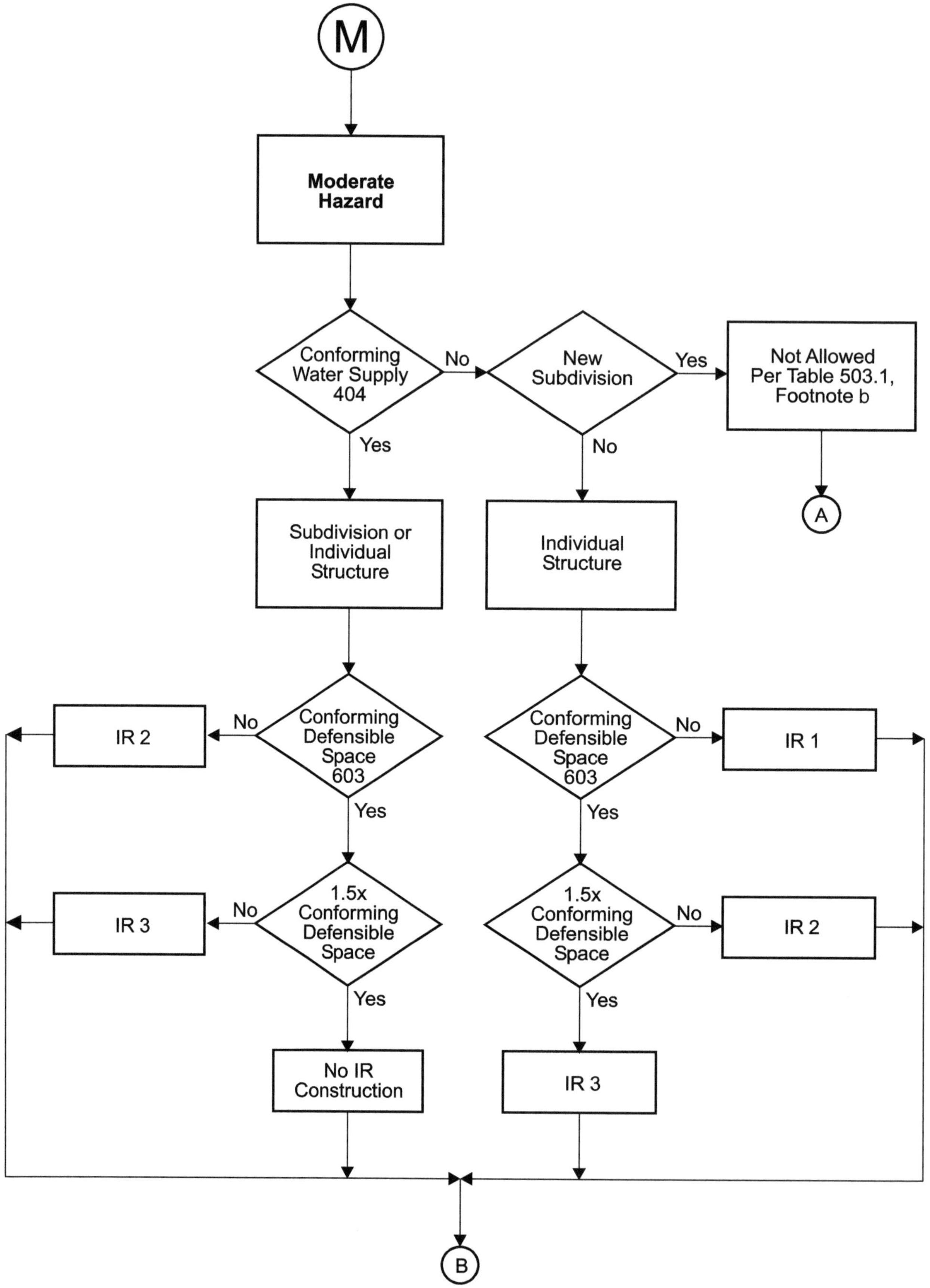

FIGURE H101.1(3)—HIGH HAZARD

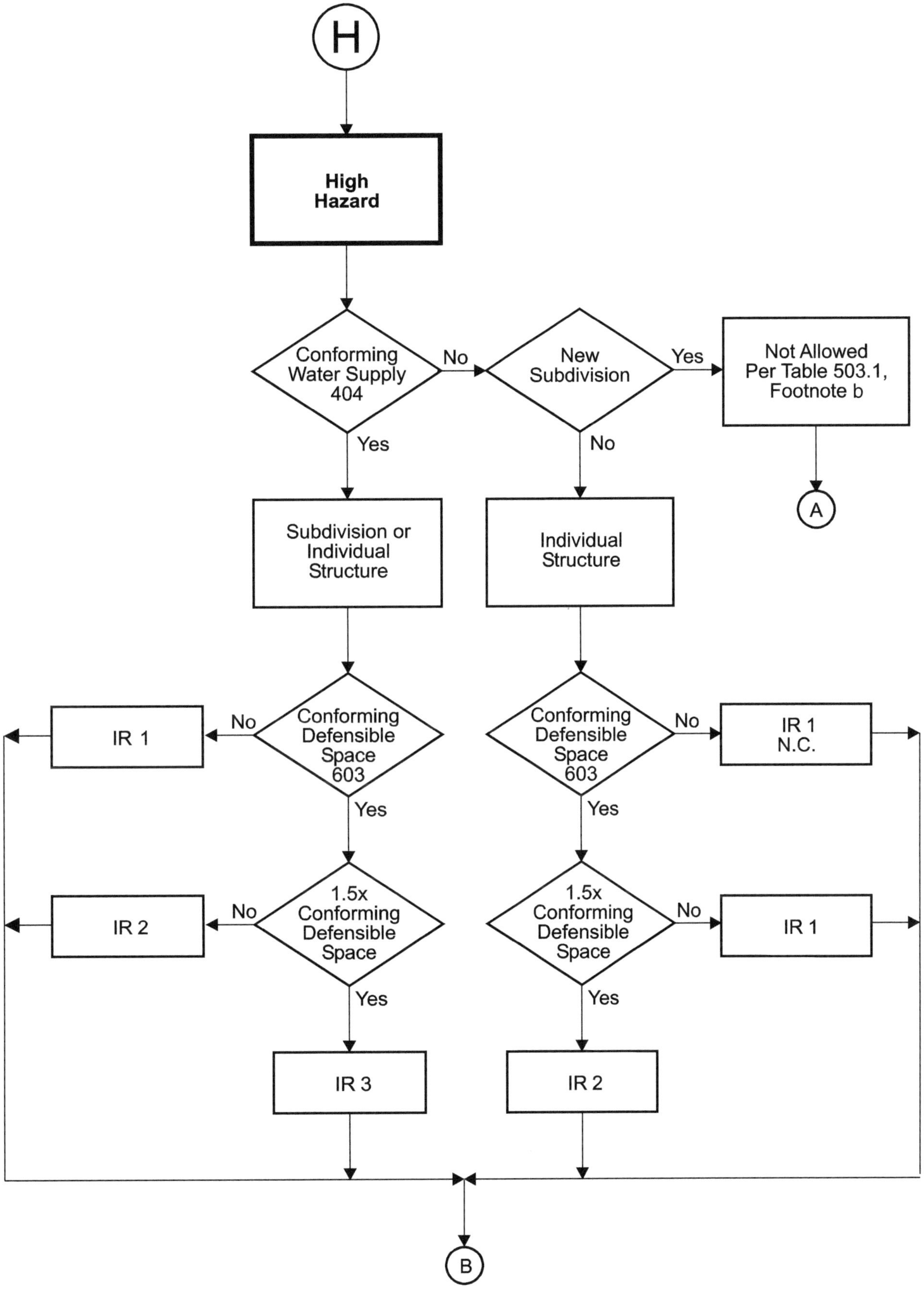

FIGURE H101.1(4)—EXTREME HAZARD

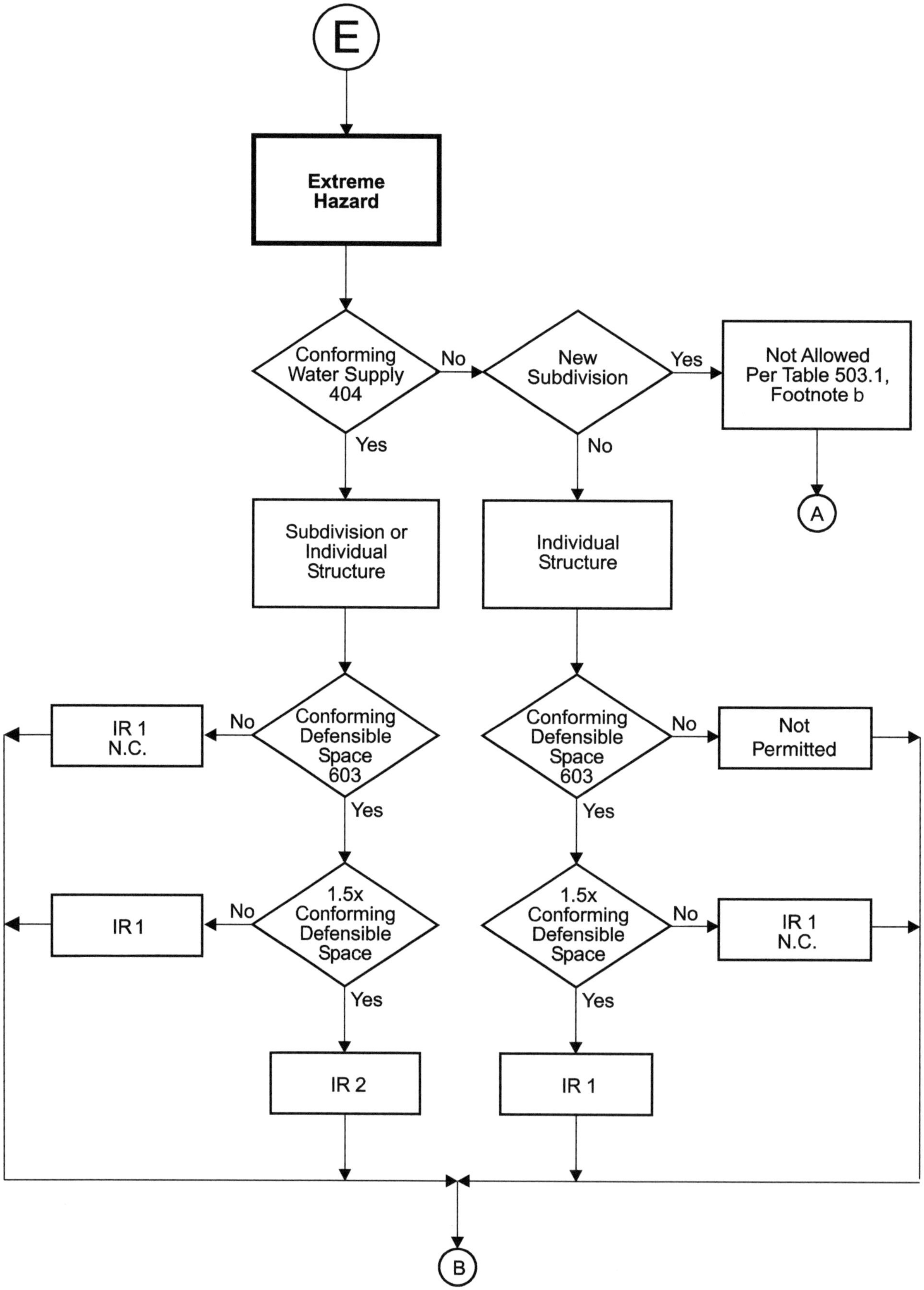

APPENDIX

I BOARD OF APPEALS

The provisions contained in this appendix are not mandatory unless specifically referenced in the adopting ordinance or legislation of the jurisdiction.

User notes:

About this appendix: *Appendix I contains optional criteria that, where adopted, provide jurisdictions with detailed appeals board member qualifications and administrative procedures to supplement the basic requirements found in Section 112.*

Code development reminder: *Code change proposals to this appendix will be considered by the Administrative Code Development Committee during the 2025 (Group B) Code Development Cycle.*

SECTION I101—GENERAL

[A] I101.1 Scope. A board of appeals shall be established within the jurisdiction for the purpose of hearing applications for modification of the requirements of this code pursuant to the provisions of Section 112, Means of Appeals. The board shall be established and operated in accordance with this section, and shall be authorized to hear evidence from appellants and the building official pertaining to the application and intent of this code for the purpose of issuing orders pursuant to these provisions.

[A] I101.2 Application for appeal. Any person shall have the right to appeal a decision of the *code official* to the board. An application for appeal shall be based on a claim that the intent of this code or the rules legally adopted hereunder have been incorrectly interpreted, the provisions of this code do not fully apply or an equally good or better form of construction is proposed. The application shall be filed on a form obtained from the *code official* within 20 days after the notice was served.

[A] I101.2.1 Limitation of authority. The board shall not have authority to waive requirements of this code or interpret the administration of this code.

[A] I101.2.2 Stays of enforcement. Appeals of notice and orders, other than Imminent Danger notices, shall stay the enforcement of the notice and order until the appeal is heard by the board.

[A] I101.3 Membership of board. The board shall consist of five voting members appointed by the chief appointing authority of the jurisdiction. Each member shall serve for [NUMBER OF YEARS] years or until a successor has been appointed. The board members' terms shall be staggered at intervals, so as to provide continuity. The *code official* shall be an ex officio member of said board but shall not vote on any matter before the board.

[A] I101.3.1 Qualifications. The board shall consist of five individuals, who are qualified by experience and training to pass on matters pertaining to building construction and are not employees of the jurisdiction.

[A] I101.3.2 Alternate members. The chief appointing authority is authorized to appoint two alternate members who shall be called by the board chairperson to hear appeals during the absence or disqualification of a member. Alternate members shall possess the qualifications required for board membership, and shall be appointed for the same term or until a successor has been appointed.

[A] I101.3.3 Vacancies. Vacancies shall be filled for an unexpired term in the same manner in which original appointments are required to be made.

[A] I101.3.4 Chairperson. The board shall annually select one of its members to serve as chairperson.

[A] I101.3.5 Secretary. The chief appointing authority shall designate a qualified clerk to serve as secretary to the board. The secretary shall file a detailed record of all proceedings, which shall set forth the reasons for the board's decision, the vote of each member, the absence of a member and any failure of a member to vote.

[A] I101.3.6 Conflict of interest. A member with any personal, professional or financial interest in a matter before the board shall declare such interest and refrain from participating in discussions, deliberations and voting on such matters.

[A] I101.3.7 Compensation of members. Compensation of members shall be determined by law.

[A] I101.3.8 Removal from the board. A member shall be removed from the board prior to the end of their terms only for cause. Any member with continued absence from regular meeting of the board may be removed at the discretion of the chief appointing authority.

[A] I101.4 Rules and procedures. The board shall establish policies and procedures necessary to carry out its duties consistent with the provisions of this code and applicable state law. The procedures shall not require compliance with strict rules of evidence, but shall mandate that only relevant information be presented.

[A] I101.5 Notice of meeting. The board shall meet upon notice from the chairperson, within 10 days of the filing of an appeal or at stated periodic intervals.

[A] I101.5.1 Open hearing. All hearings before the board shall be open to the public. The appellant, the appellant's representative, the *code official* and any person whose interests are affected shall be given an opportunity to be heard.

[A] I101.5.2 Quorum. Three members of the board shall constitute a quorum.

[A] I101.5.3 Postponed hearing. When five members are not present to hear an appeal, either the appellant or the appellant's representative shall have the right to request a postponement of the hearing.

[A] I101.6 Legal counsel. The jurisdiction shall furnish legal counsel to the board to provide members with general legal advice concerning matters before them for consideration. Members shall be represented by legal counsel at the jurisdiction's expense in all matters arising from service within the scope of their duties.

[A] I101.7 Board decision. The board shall only modify or reverse the decision of the *code official* by a concurring vote of three or more members.

[A] I101.7.1 Resolution. The decision of the board shall be by resolution. Every decision shall be promptly filed in writing in the office of the *code official* within three days and shall be open to the public for inspection. A certified copy shall be furnished to the appellant or the appellant's representative and to the *code official*.

[A] I101.7.2 Administration. The *code official* shall take immediate action in accordance with the decision of the board.

[A] I101.8 Court review. Any person, whether or not a previous party of the appeal, shall have the right to apply to the appropriate court for a writ of certiorari to correct errors of law. Application for review shall be made in the manner and time required by law following the filing of the decision in the office of the chief administrative officer.

INDEX

EDITORIAL CHANGES—SECOND PRINTING

Page 12, **Section [A] 104.2.2** line 3 now reads: such alternative is not specifically prohibited by this code and has been *approved*.

For the complete errata history of this code, please visit: https://www.iccsafe.org/errata-central/

ICC INTERNATIONAL CODE COUNCIL®

Membership

At the heart of the International Code Council's rich history is our connection with our members. The Code Council and its members work together to create safer homes, buildings and communities by promoting and maintaining the International Codes® (I-Codes®). Members benefit from top-notch professional development, innovative products and services, timely industry news and valuable networking opportunities while lending their own expertise to the code development process.

Membership Categories

Code officials, fire officials, architects, engineers, designers and building safety professionals are all valued members of the Code Council. The Code Council offers several membership categories to meet your needs.

- **Government/Municipality** – Includes agencies, departments and units engaged in administration, formulation, implementation and/or enforcement of laws, ordinances, rules/regulations relating to public health, safety and welfare.
- **Corporate** – An organization such as an association, society, testing lab, institute, company or manufacturer interested in the purposes and objectives of the Code Council.
- **Individual**
  - **Professional** – An individual who is interested in the purposes and objectives of the Code Council.
  - **Participating** – An individual who participates in and supports the vision and mission of the organization. *Participating Members do not qualify for all member benefits.*
  - **Student** – An individual who is enrolled in classes or the equivalent of at least 3 credit hours of classroom instruction per week.

Member Benefits

As a member, you can expect nothing less than world-class service and invaluable benefits!

- Unlock the I-Codes with a Digital Codes Premium subscription (based on member category).
- Receive 10–25% off code books, specialized publications and training materials.
- Take advantage of discounts on expert training and educational programs so you can earn CEUs.
- Receive discounts on certification exams, exam resource materials and credentialing renewals.
- Showcase your membership with a digital badge on social media and your resume.
- Access technical support to answer your code-related questions (based on membership category).

24-23445

Learn more about membership at iccsafe.org/about-membership

Apply today at iccsafe.org/join-icc